Hot Metal

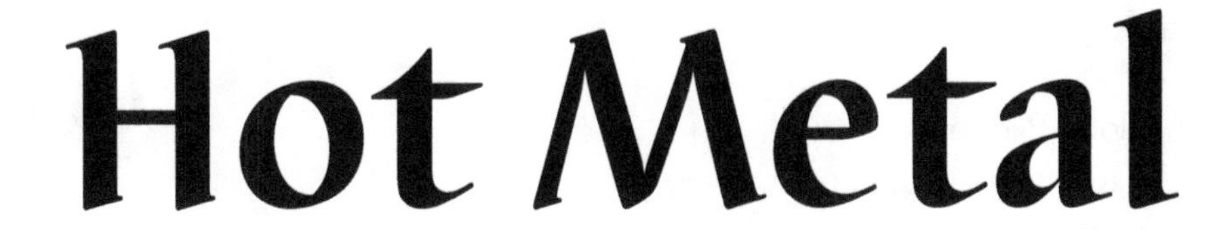

Hot Metal

A Complete Guide to the Metal Casting of Sculpture

Wayne E. Potratz

6th Edition

An Imprint of Globe Pequot
A division of the Rowman & Littlefield
Publishing Group, Inc.

An Imprint of Globe Pequot
A division of the Rowman & Littlefield
Publishing Group, Inc.
4501 Forbes Blvd., Ste. 200
Lanham, MD 20706
www.rowman.com

Distributed by NATIONAL BOOK NETWORK

All images courtesy of the author unless otherwise noted.

British Library Cataloguing in Publication Information available

Library of Congress Cataloging-in-Publication Data available

ISBN 978-1-879535-31-2 (paperback)

ISBN 978-1-879535-33-6 (electronic)

Table of Contents

Table of Contents

Acknowledgment And Author Information

Special thanks to Virginia L. Potratz, without whose love and support, my work would not be possible.

Thanks and acknowledgment to my teachers Anthony Caponi, Harold Paris, and Richard Randall. Thanks also to the following friends and colleagues: George Beasley, Bart Cuderman, Steven Daly, Chris Dashke, Kurt Dyrhaug, Thomas Gipe, Meridith Jack, Alan Kraning, William Malo, Paul McMahill, John Poole, Julius Schmidt, James Stewart, James Swartz, Cliff Prokop, and Norman Taylor.

Wayne E. Potratz is a sculptor and Professor Emeritus and Scholar of the College Emeritus of the University of Minnesota, Twin Cities, where he did creative research in metal casting and taught sculpture from 1969–2014. His studio is in Minneapolis; his work can be seen at www.ironwain.com

"The artist is nothing without the gift, but the gift is nothing without work."

— Emile Zola

"Where the mind goes, the tongue does not necessarily follow."

— Unknown

"First learn to be a craftsman; it won't keep you from being a genius."

— Eugene Delacroix

Metal Casting – An Introduction

Chapter 1

"The nobility of an art depends on the purity of the desire that gives rise to it, and the artist's uncertainty as to the happy outcome of his activity. The more uncertain he is of the results of his efforts, on account of the nature of the material he wrestles with and the means he employs for subjugating it, the purer his desire and the more evident his worth.

Among all the arts, I know of none more hazardous, none less certain of the outcome and consequently more noble, than those which call for the use of *Fire.* By their nature they exclude or punish any negligence; allow no relaxation or respite, no fluctuations of thought, courage, or temper. They emphasize, in its most dramatic aspect, the close combat between man and form. *Fire,* their essential agent, is also their greatest enemy... All the fire worker's admirable vigilance and all the foresight learned from experience... still leave immense scope for the noble element of uncertainty. They can never abolish *Chance.* Risk remains the dominating and, as it were, the sanctifying element of this great art."

— Paul Valéry, "On the Preeminent Dignity of the Arts of Fire"

Welcome to the discipline of the metal casting of sculpture. By practicing this art technology, you are participating in a human tradition that is at least 8,000 years old. Your task, and that of all of us who choose to follow this way, is to expand this tradition in a context of reverence and respect while seeking to make innovations and new imagery. This tradition makes use of the five essential elements: Earth, Air, Fire, Water, and Light.

Earth Metals are the body of the earth: iron is the earth's flesh, copper is the earth's blood, clay and sand are the bones of the earth and the "mother of the casting." The earth also is the source of the fuels that melt metal. These materials, from "our Mother Earth," are not unlimited. They should be used with care, conservation, and respect. It is not surprising that metals have been associated with ritual and magic tradition—the transformation of metals into sculptural form is indeed a magical process.

Air The air which combusts the fuel to make the fire which melts the metal is also the air of your respiration—the breathing in and the breathing out of action, that reminds us that the mind extends throughout the body. The whole body is also the whole mind. Air is the breath of life for the artist and the artist's creation in metal.

Fire Fire is a useful and respected friend, but it can also be a dangerous and deadly enemy. To do metal casting is to understand, respect, and befriend fire. Fire is the catalyst of air and earth that transforms solid metal to liquid metal enabling the mold to do its work. Fire is the force of chance in the fire arts. Fire warms our body while also providing the spark of imagination to light our vision.

Water As the antithesis of fire, water is the catalyst which makes the molding of form possible. As the material of life, it makes up most of your body and carries with it the process of growth and discovery, balancing the fire in the harmony of life.

Light Light from the material realm of earth, air, fire, and water illuminates form and reveals the concepts embedded in the form (vision), while *light* from the metaphysical realm forms and nourishes those concepts (vision). Both lights reveal the weightless and metaphysical aspect of art . . . "Where there is no vision, the people perish"— Proverbs 29:18

In the foundry, we respectfully dance with all these materials and transform them according to our vision and our *vision*. Do not disregard any of these materials, get to know them and treat them as trusted friends, think none of them as being insignificant and unimportant, cast none of them aside.

This tradition uses *tools.* Tools are an extension of your self. They are the helpmates in the process. They enable and they give us power. They are the "Kraft"—the power to make your *vision* visible. To abuse and misuse tools is to abuse and diminish your own strength, and because of the collective nature of the foundry, the abuse of tools, abuses and diminishes the collective users, since we all share in the power.

This tradition expands and honors *process.* Process is the collective experience of a tradition and the methodology whereby material is transformed by concept into *art.* To disregard process results in the waste and dishonoring of materials, the abuse of tools, and the production of "cast-aways" rather than "castings." One should follow on this path of process with both eyes on the path and the third eye on the horizon of innovation. One neither blindly follows this tradition (the craftsman) nor blindly ignores this tradition (the fool), but rather walks along the path with an awareness of where it has come from and where it might lead (the Artist).

This tradition works through *enantiamorphs.* As molds, enantiamorphs are the three-dimensional mirror opposites of your formal and conceptual desires. They are the space that is "To Be." They are the inverse relationships of what "Can Be." Just as space and time are the inverse of each other, yet relate to form the "now" which we experience as reality, the mold and the metal are the inverse of each other in the form of art which we take as relevant and necessary. The anticipation of the consummation of this relationship in the moment before the liquid metal is poured into the enantiamorphic space of the mold, accounts for the existential angst you will feel. This feeling is both the fear and the joy of what might be. Be kind to you mold!

This tradition combines dance and magic, ritual and knowledge, action and thought, strength and guile to make an art that is always changing, yet always the same. For it is precisely *art,* which differentiates human beings from all other life forms on this planet, makes us truly human, and connects us to the spiritual realm. Welcome to the path—may your journey be filled with discovery.

"Nor do I doubt that whoever considers this art well will fail to recognize a certain brutishness in it, for the founder is always like a chimney sweep, covered with charcoal and distasteful sooty smoke, his clothing dusty and half burned by the fire, his hands and face all plastered with soft muddy earth. To this is added the fact that for this work a violent and continuous straining of all a man's strength is required which brings great harm to his body and holds many definite dangers to his life. In addition, this art holds the mind of the artificer in suspense and fear regarding its outcome and keeps his spirit disturbed and continually anxious. For this reason they are called fanatics and are despised as fools. But, with all of this, it is a profitable and skillful art and in a large part delightful."

— Vannoccio Biringuccio, *Pirotechnia* [1540]

Recommended Tools and Equipment

Chapter 2

For sand molding

1. Assorted 1/4" shank mounted abrasive stones
2. Assorted old drill bits or masonry bits
3. Used 1" wide hacksaw blade for sawing sand and scoring part lines
4. Used carborundum wheels for dressing molds
5. Cotton bag or old cotton sock for parting compound
6. Ketchup or honey squeeze bottle for core paste
7. Electric drill
8. Paint mixer with 1/4" shank
9. 1" natural bristle brush
10. Coffee can with lid or jar with lid for mixing mold wash
11. Claw hammer, old screwdriver
12. Slick and spoon
13. Toolbox for sand tools

For wax working

1. Coffee cans or old kettles for wax mixtures
2. Sharp pocket knife or carpet knife
3. Cheap electric soldering iron or old-style soldering iron
4. Old table knives, grapefruit knife, painting knife (thin blades)
5. 1" wide putty knife
6. Cheap 1" wide natural bristle brushes for painting hot wax
7. Meat thermometer
8. Old pantyhose or sandpaper for smoothing wax
9. Mineral spirits
10. Used dental tools
11. Cloth or leather gloves

For metal chasing and finishing

1. Ball-peen hammer
2. Several files with handles (half-round bastard file recommended)
3. Cold chisels (1/4"–1" wide)
4. Chasing tools made from old files
5. Rotary files (1/4" shank)
6. Emery cloth, 5" sanding disks to fit hand-held grinder
7. Flap wheels and grinding wheels

For patina

1. Plastic spray bottles with spray nozzles
2. Glass or plastic jars with lids for storing and mixing patinas
3. Old tablespoon
4. Funnel

A Brief History of Metalcasting

Chapter 3

Perhaps the first person to discover cast metal was a woman potter tending a kiln near a site rich with copper and tin ores. She had used the ores thickly as rough glazes on her pots. In the morning, while raking away the ash from the firing, she found a bright button of metal on the floor of the kiln near her vessels. The Bronze Age was born.

The earliest evidence of metal casting activity now dates from before 5,000 BCE in the Carpathian Mountain area of Eurasia. The development of ceramic pottery and the ability to master fire in kilns transferred easily to the rise of metal casting technology. From Eurasia it spread with amazing speed as the superiority of tools and weapons made with copper and bronze enabled the human imagination to begin to build civilized societies. The discovery of the eutectic property of tin produced the first bronze—harder than copper, but with a lower melting point. Other alloying metals such as lead and zinc brought new properties. Bronze helped forge the great civilizations of Egypt and Mesopotamia and helped usher in the twin civilizers of agriculture and animal husbandry. In China, as early as 1700 BCE., during the Shang dynasty, metal casters were making works of incredible intricacy, amazing thinness (less than 4–6 mm thick), and great beauty. In the history of metal casting, the theory that aesthetics is the driving force of technological and social development makes itself clear.

The gradual development of sand molding coupled with the demand for cast iron firebacks, cast iron pipe, cannons, and the development of the stove, spurred technological advances in cast iron. The box stove was invented in Germany that coincidentally was also an area known for fine woodcuts. Woodcut dies were used in the ornamentation of firebacks and stove parts and developed into the wooden pattern, used so extensively in sand casting. Metal patterns were also used at this time. The intricate and delicate castings in iron produced in Berlin between 1776 and 1825 are wonderful examples of cast iron produced from these wooden patterns. The Lamprecht Collection in the Birmingham Museum of Art [Alabama] has many examples of these cast iron works.

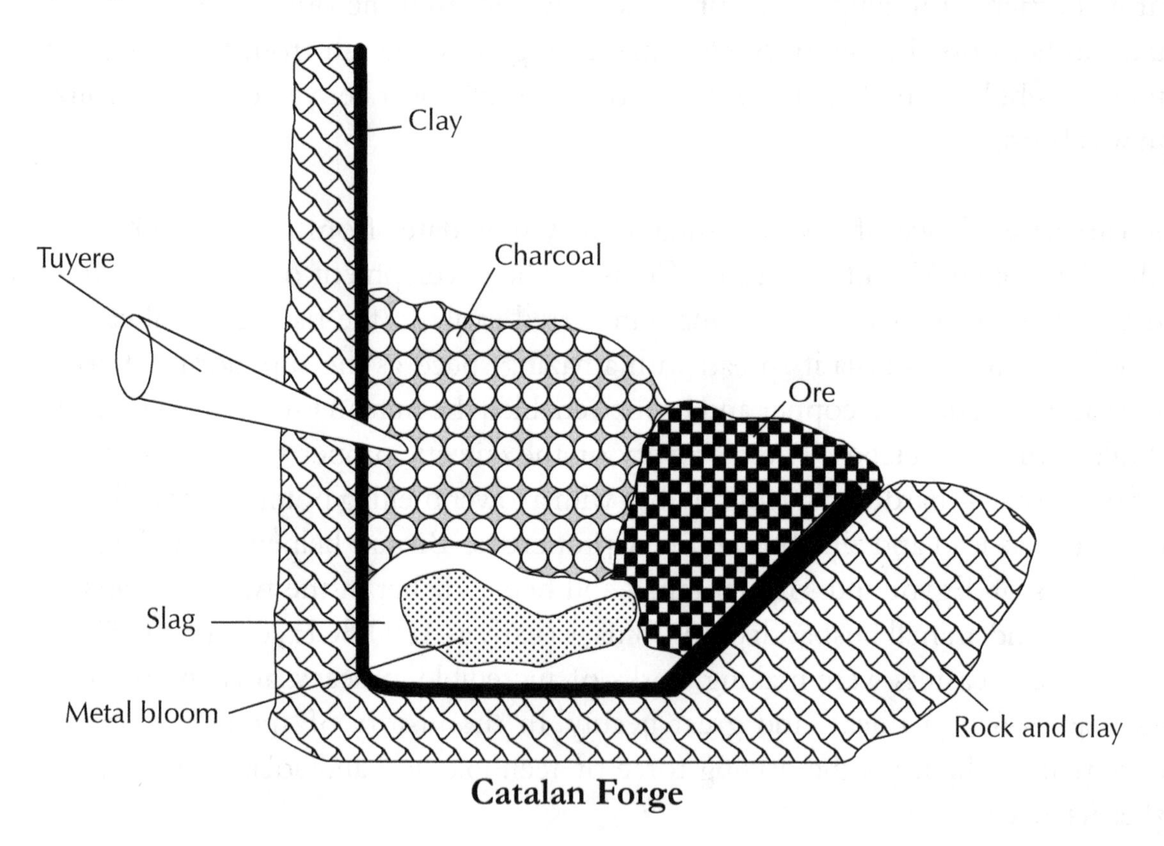

Catalan Forge

Early castings were made in permanent stone molds, clay molds, and also in sand. Many examples of stone molds can be found in museums; knives, axe-heads, spear points, and other tools are the predominant forms found cast from these molds. The first furnaces were most likely built on the sides of hills to take advantage of the wind. From primitive clay lined pits to standing furnaces made of clay and fired with the aid of animal skin bellows and tuyeres (openings from the air source to the interior of the furnace) at the firebox level, furnace development progressed steadily. By 2000 BCE furnaces were hot enough to smelt iron. As with other metals, iron, in its native form (meteorite iron) and in its earliest smeltings, was regarded as a precious object.

Around 500 BCE, due to the development of double chambered bellows, the Chinese were casting iron. The box bellows, made of wood, enabled the Chinese to produce a steady stream of air, since it "blew" on both the push and the pull stroke. This invention coupled with the discovery that when iron is heated and held at a high heat in contact with carbon, its melting point drops from 2786° F [1530° C] to 2138° F [1170° C], enabled the Chinese to take the forefront of metal casting technology. Chinese advances in metallurgy produced a high phosphorous iron (6–7% P) that is very fluid and can be cast at 1796° F [980° C] that is below the melting point of bronze. The presence of low sulfur coal near iron ore deposits in China also contributed to the rapid development of cast iron. Cast iron objects became common household items in China centuries before iron was cast in Europe. Iron was even used architecturally to roof pagodas! The use of iron sculpturally is dated as early as 502 BCE. A large cast lion in Hopei province was cast about 953 CE. It is 16 feet long, 20 feet high, and has a metal thickness ranging from 2"–8" [5–20 cm.].

In Japan, the use of bronze dates from the introduction of Buddhism from China around 500 CE. Many large scale castings also exist in Japan. The great temple bell at Nara is 13 feet high and 9 feet in diameter. Also in Nara at the Todaji Temple is the Diabutsu of Nara, first cast in 752 CE. This colossal sculpture is 52 feet high [15 meters] without pedestal and weighs 250 tons [226,796 kilos]. At Kamakura, another Diabutsu was cast around 1252 CE. This huge bronze sculpture of the seated Buddha is 97 feet in circumference and 42 feet high [Roughly 11.3 meters in height; 13.35 meters with pedestal] and weighs 125 tons [113,398 kilos].

Japanese cast iron, especially teapots for the tea ceremony, achieved a high level of aesthetic beauty. Teapots are traditionally cast with a method called "Sogata" which employs clay molds and metal sweeps to produce the form.

The Japanese "tatara" furnace is a peculiar kind of rectangular, low shaft blast furnace which produced both pig iron and the high-carbon steel, known as "Tama-hagane" used to forge Japanese swords. This furnace is made of fire clay and has two rows of 18–20 tuyeres on each side. The furnace is supported by an elaborate foundation more than 6 meters deep that was completely insulated and moisture free. The tatara is fired with charcoal and magnetic iron sand is melted to obtain both the iron and the steel. It takes 72 straight hours to produce 2,250 kg. of pig iron or a 2,900 kg. bloom of wrought steel. The tatara furnace is still being operated in the village of Izumo Yakota, Shimane Prefecture ["NITTHO Tatara", a division of the Hitachi Steel Company] and is unique in its ability to produce high quality steel directly from ore.

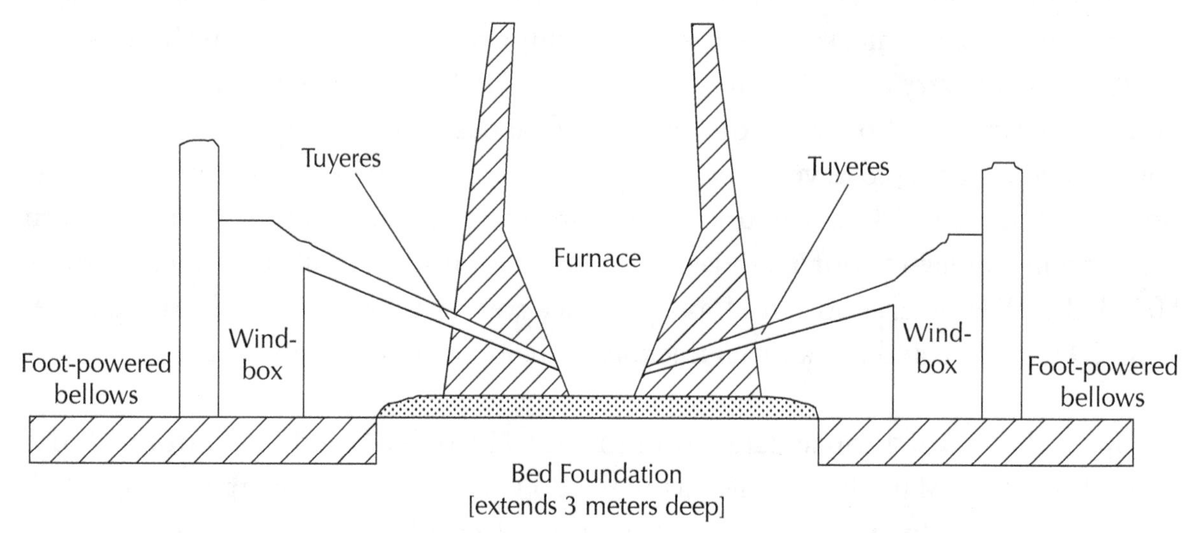

Japanese Tatara Furnace

In India, metalcasting of copper and bronze was well known by 3000 BCE. The lost wax casting of figures in copper, tin, gold, silver, and lead also flourished. While iron forging is recorded as early as 1000 BCE., iron casting lagged behind the Chinese. The Indians did, however, develop cementation or "blister" steel by heating wrought iron in a charcoal fire and also produced the first cast crucible steel around 500 CE. However, this process of "wootz" steel production became extinct and it was not until the 18th century. that a reproducible cast steel process was developed.

The development of foundry practice seems to go hand in hand with the development of ceramics. Both crucibles and clay molds depended upon the potter's technology. Egyptian tomb paintings clearly depict the production of crucibles and the casting process. Other groups such as the Phoenicians spread the technology of cast metals through their extended trading network and commerce in metal ores. In fact, the legendary Hiram of Tyre, was a Phoenician founder whose works for Solomon are the first castings connected with a historical figure. "King Solomon's Pool," cast about 1066 BCE, was a bronze bowl 15 feet in diameter and 7.5 feet high [4.5 meters deep and 2.3 meters high] and stood on eight cast bronze oxen. Hiram of Tyre also cast other works for Solomon including the massive pillars for a temple in the Jordan River valley, but no trace of these works, other than the historical record, exists today.

The presence of copper ore on Cyprus gave rise to the metal casting traditions of the Minoans and Greeks. In fact the word "copper" derives from "Cyprium," which the Romans called "coprum." The beginnings of alchemy produced other alloys of bronze as zinc was alloyed with copper to produce the first brasses.

The Greeks developed techniques for casting bronze figures, many of which can be seen today in museums; drawings depicting Greek foundry practice can be seen on Greek ceramic vessels. The most famous casting of antiquity was the Colossus of Rhodes. This 105-foot-high [32 meters] sculpture of Apollo was cast and erected around 290 BCE. The figure held a torch, which served as a beacon as it straddled the harbor at Rhodes. The figure was destroyed by an earthquake in 224 BCE and, like so many cast works, was no doubt, remelted into the form of other works.

The Greeks also advanced the metallurgy of iron. By 850 BCE, they had developed a technique for tempering iron by heating it and quenching it in oil. There is evidence that the Greeks knew how to cast iron through the works of Theodomus the Samian, who is credited with casting an iron throne and two figure sculptures in iron.

While the Roman author, Pliny the Elder, describes something of the metallurgy of his time, it wasn't until the 11th century, when the monk Theophilus wrote *Schedula Diversarium Artium*, that foundry information was codified in a text. It was not until the 16th century that anything else about metal casting technology was written down when texts by Biringuccio and Agricola (see *Bibliography*) appeared.

The so-called Dark Ages and the Middle Ages in Europe were marked by slow developments in furnace technology for the production of iron. From the early Catalan furnace to the German furnaces and then to furnaces such as the Swiss furnace and the Swedish Osmond furnace, higher and higher temperatures began to be achieved. With the invention of the water-driven bellows system around 1300 ACE and taller furnaces, the production of molten iron directly from a blast furnace was realized. The fuel for these furnaces was charcoal, and furnace sites were often located in areas where ore and forest was plentiful. From 1300 to 1700 ACE, cast iron was, in the most part, produced directly from ore in these large blast furnaces.

The Renaissance sparked a revival in the casting of bronze sculpture. Many of the sculptors were also founders. The great bronze doors of the Baptistery in Florence by Lorenzo Ghiberti were cast during this time. Young Leonardo da Vinci, as an apprentice of Verrocchio, is reputed to have worked on the large horse and rider in Venice, known as the "Colleoni Monument." For an interesting (and amusing) tale of the hazards of metal casting in the 1500s, read "Vita" by Benvenuto Cellini, which describes, among many other aspects of Renaissance life, the casting of the "Perseus" in 1540 ACE. Whether cast in loam molds or with the lost wax process, these sculptures stand as some of the finest expressions of both technological and aesthetic achievement.

The destruction of large tracts of European forests to fuel the increasing demand for iron, spurred the search for an alternative fuel. Around 1619, Dud Dudley, an English iron founder invented a furnace that was fired with coal, but he was driven out of business by the charcoal mongers. He wrote a book on his procedures in 1665 called *Metallum Martis*, which influenced later iron founders. Abraham Darby of Coalbrookdale, England, was the first to reduce coal to produce coke (made in the same manner as wood is reduced to make charcoal) and built the first coking ovens. His ability to produce iron at much less cost than the charcoal melters changed the face of the foundry industry and helped usher in the "Industrial Revolution." Darby is also credited with inventing the foundry flask, which freed founders from having to mold in pits or on the floor. The famous "Iron Bridge" near Telford, Shropshire, England, was made possible by the Darby's developments and was the first structural use of cast iron.

The 18th and 19th centuries brought tremendous advancements: From the French remelt furnaces and the annealing process invented by Rene Antoine de Reaumur to the development of the cupola by John Wilkinson in 1794 and the rediscovery of crucible steel by Benjamin Huntsman, the technology of metal casting matched the tremendous industrial growth of the period. These advances aided the development of the steam engine and the railroad and a myriad of other machines and inventions that forever changed the face of human civilization.

The great Meso-American civilizations of South and Central America also produced metal castings. Gold and silver were the predominant metals cast; there's no evidence of iron casting, but cast bronze artifacts, made using wax patterns and cast in clay/charcoal molds have been found, especially in Panama. Copper, traded from sources of native copper around the Great Lakes, is also found in many North American and Central American indigenous cultures.

The Saugus Iron Works near Lynn, Massachusetts, founded in 1642, was the first foundry in colonial America. The first casting from this works is known as the "Saugus pot" and is a small cooking pot with tripod legs. A typical 18th-century "iron plantation" is preserved by the National Parks Service (see Bibliography) at Hopewell Village near Birdsboro, Pennsylvania. This area near Valley Forge is rich in American casting history and was the source of much of the colonial arsenal during the American Revolutionary War.

By the late 19th century, iron was used for everything from toys to architectural facades, while bronze and brass predominated in sculpture and decorative objects. Aluminum had been smelted from bauxite ore, but was not yet a metal of importance. Further refinements in melting techniques such as Bessemer's converter for steel production, open hearth furnaces, and carbon arc electric furnace contributed to the industrial growth into the 20th century. Today high-frequency induction furnaces seem to be the predominant way that massive amounts of cast metal are produced. For the sculptor, the 20th century with its shift from naturalistic sculpture to abstraction brought a move away from such traditional materials as bronze. Although sculpture in bronze continued to be made, its use by "avant garde" sculptors in the 1950s, 60s, and 70s greatly diminished. Iron, with its strong connection to the industrial image, was never a particularly popular metal for sculpture. In the 1960s, a revival of metal casting by artists who wanted to control their own casting began in universities and art schools. As a result, several generations of sculptors with extensive metal casting ability are now working in cast bronze, cast aluminum, and cast iron; many have supported themselves by developing casting facilities and casting the work of other artists. The present mood in the art world seems to be a return to traditional materials such as stone and bronze. The durability of cast metal, its ability to assume such a wide variety of forms, and its potential for beautiful color makes it a material which artists will use far, far into the future. Artists have not only adopted and adapted technological developments, but, as in the past, aesthetic innovation will also bring technological innovation. Your own imagination and creativity may be the key to just such a development and change.

SAFETY AND RIGHT TO KNOW

Chapter 4

Safety First! The foundry is a potentially dangerous place. This discussion is intended to familiarize you with the potential hazards and risks so that your experience will be an enjoyable and healthy one. The key to a healthy and profitable experience is the knowledge of correct technical procedures and an enlightened adherence to those procedures. This is not intended to discourage experimentation, but because the foundry is often a communal studio any violation of correct technical procedure to produce an aesthetic effect should be discussed with an experienced founder prior to any experimentation.

More than in any other artistic discipline, you will become aware of the communal nature of the foundry studio—it is simply impossible to do it by yourself, if your work is large in size. Becoming aware of how your actions and habits affect others is a good way to participate in a safe environment. Everything in the foundry requires your cooperation with your fellow artists. It is essential that you not only be aware of the hazards to yourself, but that you take responsibility for

Golden Rules

1 If you open it, close it.
2. If you turn it on, turn it off.
3. If you unlock it, lock it up.
4. If you break it, admit it.
5. If you can't fix it, call someone who can.
6. If you borrow it, return it promptly.
7. If you value it, take care of it.
8. If you make a mess, clean it up.
9. If you move it, put it back.
10. If it belongs to someone else, get permission to use it.
11. If you don't know how to operate it, leave it alone.

others, especially if you become aware of somebody putting themselves in jeopardy. Likewise it is essential to lend a helping hand (even though you may not be asked). You will soon be needing help yourself.

Participation in any foundry activity requires the following safety equipment:

1. Long-sleeved wool or close-weaved cotton shirt or a cotton or leather jacket. (Do not wear synthetic fabrics as they will melt if metal hits them.)
2. Leather boots or sturdy high-topped shoes. Steel-toed shoes are recommended for serious sculptors. **Do not wear sandals or athletic shoes in the foundry.** Consider "kick-off" leather shoes that do not have laces.
3. Safety glasses with side spark screens. Welding supply stores and safety equipment stores stock safety glasses.
4. Leather gloves with gauntlets or kevlar gloves. Gloves should be at least mid forearm in length. Leather gloves tend to shrink in the palms when handling hot objects and kevlar gloves do not stand up well when handling rough materials. You may want to have a cheap pair of leather gloves for mold handling and save the kevlar gloves for hot work only.
5. Cap or hat (not synthetic material) to cover your head and to keep your hair tied back. A heavy cotton kerchief or scarf will also work.
6. Loose-legged cotton pants. Do not wear tight jeans in the foundry, as the radiant heat will burn right through the cloth that is tight against your skin.
7. Face shield (metal mesh face shield or melt resistant clear plastic) and hard hat.
8. Leather apron and leather spats that cover the lower legs and ankles.
9. Dust mask (Bureau of Mines approved for dusts) with two straps.
10. If you are spending a lot of time around the furnace or are working in poorly ventilated areas, you may want to consider a filtered helmet such as the Racal "Airstream."

Some materials, supplies, and equipment are potentially hazardous to your health if the proper precautions are not observed. You should know and follow the correct procedures in dealing with the following materials and equipment:

1) Dusts - Silica (silica sand, silica flour, ceramic shell slurry, ceramic shell stucco), zircon (mold wash), graphite (also called "plumbago," mold wash and parting compound), talc (N.S.P.- non-silica parting compound), clay (western bentonite, ball clay), plaster, and wood flour. These materials are hazardous to your lungs below a 10-micron particle size. Use of your dust mask is required when handling or mixing these materials. This includes digging sand from a pit, mulling sand to make molds, sweeping, working with ceramic shell, parting molds, mixing mold washes, and mixing plaster. Avoid getting these materials into the air through vigorous mixing, shoveling, or sweeping. Clean up all spills promptly. Proper ventilation systems are recommended and must be on when using these materials.

2) Patina Chemicals - Many of the chemicals used in patinas are toxic if inhaled or swallowed, and can be hazardous to your skin or eyes. When mixing or applying patina formulas, wear disposable gloves, eye protection, and your dust mask. Avoid spilling chemicals and clean up all spills promptly. Return all chemicals to their proper storage area. When mixing acids, **always add acid to water, never do the reverse** as it will cause an explosion. Be careful to keep oxidizers (such as ferric nitrate) from coming into contact with organic materials as this is a fire hazard. Label all containers of patina mixtures with their contents, date mixed, and the name of the person mixing the patina. Not only is this an OSHA requirement, but it easily identifies the formulas and their effects. When applying patinas, use a hooded and ventilated patina area. Sweep and wash tables and other surfaces thoroughly when finished in any patina area.

3) Pouring Metal -

A. General considerations: Remember that kiln firings and the flashout of shells must be adjusted to the pouring dates and times. It's a good idea to chalk the names of the persons involved, the number of molds, metals to be poured, and the estimated weights to be poured on a blackboard prior to the pour (this facilitates organization of the pour and the number of heats needed). **Safety glasses and safety gear must be worn in the furnace and pouring areas whenever the furnaces are in operation or metal is being poured.** Safety glasses must also be worn during any chasing, grinding, sanding, cutting, or drilling operations. "Blind artists are exceedingly rare."

Remember: It's a good idea to clean up the foundry area immediately after a pour, returning all unused metal (cups, sprues, vents, spills) to the appropriate bin so that different metals do not cross contaminate one another. Miscasts and other returned metal should be cut up into pieces smaller than 4 inches, so that they will easily fit back into the crucible for remelting.

B. Metal fumes and fumes from combustion: The melting of bronze and aluminum in gas-fired furnaces, charcoal or coke fueled furnaces, induction furnaces, and the melting of iron in coke-fired cupolas or cupolettes results in fumes from both fuel combustion and fuming of the liquid metals. Bronzes which are alloyed with zinc can release zinc oxides if overheated, resulting in "metal fume fever" or "welders fever" if inhaled. Some brasses can release lead or beryllium fumes that are highly toxic. Combustion of coke produces carbon monoxide and other fumes, such as sulfur, from the burning coke. **Be sure all ventilation systems (overhead and hoods over the furnaces) are operating during melting.** Use auxiliary fans during long pours. Avoid standing directly over unventilated crucibles. Never plunge degassers or grain refiners unless the crucible is under a hood (for example, some degassers for aluminum release chlorine gas when plunged in the molten metal), or place your furnace outside if possible.

C. Fire and heat: **Be sure that protective jackets, pants, spats, face shields, safety glasses, hats and gloves are on properly before feeding metal into the crucible or performing any pouring operation.** Altering safety equipment (using duct tape, for example) renders any warranty invalid. Gloves should be tucked underneath the jacket to avoid the chance of hot metal jumping into the glove. Assume all equipment, molds, and spilled metal are hot! Wear your gloves after the pour when cleaning up. Remember that radiant heat can burn you at a distance; even those merely observing a pour should be wearing glasses and gloves. Kevlar protective sleeves that fit over jackets also give added protection.

4) Binders for Moldmaking -

A. Resin-bonded sand: The catalyst for alkyd resin binders is an isocyanate and can cause skin and lung irritation. When handling both the resin and the catalyst, wear disposable latex gloves and a fume mask. For furan binders, hardening catalysts are sulfuric acids, sulfonic acids, or phosphoric acids. Avoid spilling both materials and wipe up any spills promptly. Throw the catalyst measuring container into the trash after use. For further information, consult the Material Safety Data sheets available for the binders you are using.

B. Sodium silicate bonded sand: Sodium silicate is a caustic material that can dry skin or cause burns with longer exposure. Use disposable gloves when handling and mixing this binder. Wipe up spills promptly. For further information, consult the Material Safety Data sheets.

C. Ceramic shell slurry: The slurry is slightly caustic and will dry your skin. Use disposable gloves when dipping shells. Avoid dripping the slurry on the floor and table surfaces, as it dries and becomes powdery, it is a dust hazard (see no.1 above). For further information, consult the Material Safety Data sheets.

5) Bridge Cranes and Hoists - Read all instructions on the control handle before operating. Check to be sure that all cables, chains, or slings are not frayed and are undamaged prior to use. Be sure the spring-loaded safety-snap on the hook is operating properly before attaching cables, chains, or slings. Never lift any load over your head or get under any load. Be sure heavy loads are under control when moving the crane so that they do not swing into you or others; ask for help in controlling the load.

6) Induction Furnaces - Induction furnaces emit high frequency waves that may interfere with medical devices such as pacemakers. Persons with such devices should not be allowed in the foundry whenever the induction furnace is in operation. Warning signs should posted on the entry doors door whenever an induction furnace is in operation. Tilting induction furnaces pose other hazards, such as bridging, charging wet metal causing explosions, electrocution hazards, and the potential for the crucibles to crack resulting in metal seeping into the cooling coils (another explosive situation). Operators of induction furnaces should thoroughly understand their operation and follow all safety instructions from the manufacturer.

7) Mold Washes - Mold washes usually contain zircon or graphite mixed with isopropyl alcohol. Store mixed washes in a closed container. Both the alcohol and the fine zircon and graphite are lung hazards, therefore, all spraying of mold washes should be done in a well-ventilated area. Mold washes may also be painted on or flooded on. If you are lighting the mold wash to eliminate the alcohol vehicle, be aware of the fire hazard. Using too much alcohol or too much mold wash may also affect the binder strength if the mold wash burns for a long period.

8) Wax - Hot wax causes deep burns because you cannot get it off your skin quickly. Wear glasses, gloves, and long sleeves when pouring wax into molds or batts. Be careful of burns from wax tools; remove tools and turn off burners when leaving the wax area. Never place a wax pan with cold solid wax directly on the burner. Set it off to the side so the wax melts up the side of the pan first, then set directly on the burner to complete melting. Do not leave wax pans unattended. The flash point of the microcrystalline wax is 560° F [293° C]. Be sure that an overhead ventilation system is on when melting wax. Do not overheat wax. Overheating wax not only causes petrochemical fumes, but destroys the plastic quality of the wax. Avoid getting water in your wax, because during the melting process, the water boils and creates an explosive situation. Electric wax pots work well and avoid the possibility that open flames will ignite the wax. Old electric crock pots are a cheap alternative.

9) Heavy Weights - Be careful lifting mold weights, ingot molds, ingots, sand barrels, and molds. Always lift with your legs and keep your back straight. Use a two-wheeled cart to move weights and ingots over longer distances. Make sure you have a firm grip on weights, ingots, and ingot molds so that they do not fall on your feet. Investing in steel-toed shoes is a good idea. Molds weighing more than 75 pounds [34 kilos] should have lifting devices in the molds so that they can be moved with a hoist (see molding section).

10) Noise - You should protect your hearing with earplugs whenever you are using air die grinders or electric sanders. You may also want to use ear plugs during extended pours as the roar of the furnaces and ventilators can be annoying.

11) Air Gun - Never hold the air gun against your own or another's skin and release air. The pressure can be sufficient to blow air right through your skin causing a stroke or pulmonary embolism!

Remember: Safety First!

Pattern Making for Sculpture

Chapter 5

By definition, metal casting means the use of temporary forms that are used to cast a permanent form in aluminum, bronze, iron, or other metal. What is most important in sculpture is the three-dimensional realization of your concept. The type of pattern material is really a function of your concept. With experience, you will easily be able to select the proper material for your idea and vision. In this chapter, pattern materials that are "lost" such as wax and expanded polystyrene, as well as pattern materials which are not destroyed in the mold-making process, such as wood, plasticene, plastics, plaster, and other materials, will be discussed.

Wax has been used for centuries for the casting of metal sculpture and as a "sketching" material. In the Renaissance, wax was used to create models of scenes from which painters worked to realize the finished composition. Many museums have examples of sculptures in wax by well-known artists, many of them retrieved from the artists' studios as models and sketches rather than finished works. Historically, beeswax was the wax available for modeling. Today there are hundreds of waxes: waxes for modeling, waxes for casting in molds, waxes for carving and construction, and waxes for injection into permanent molds.

Most sculptors use a microcrystalline wax because it is similar to beeswax and, because it is a by-product of the oil industry, it is considerably less expensive to use. Various microcrystalline wax mixtures can be made to adapt to a sculptor's various requirements, such as modeling, casting, or construction. The following materials can be added to a basic wax such as the microcrystalline Victory Brown #155 made by the Bareco Wax Company:

1. Bees wax, melting point (M.P.) 144–147° F [62–64° C], very plastic and tough, used to lower M.P. and make modeling mixes more plastic.
2. Carnuba, a vegetable wax, M.P. 183–189° F [84–87° C] a very brittle and hard wax used to raise M.P. and harden waxes for warm temperature work, has the property of picking up fine detail when used in molds; often used in first coat mixtures in delicate molds.
3. Paraffin, M.P. 118–144° F [48–62° C], a cheap wax, not very plastic, shrinks a great deal upon cooling; a good wax for stiffening back-up layers when pouring wax in molds.
4. Common resin, a residue of turpentine, M.P. 160–176° F [71–80° C], toughens wax mixtures, extends modeling time, and smooths out pouring mixtures (used up to 50% of the mix; tends to precipitate out of mix in smaller percentages).
5. Gum dammar, M.P. 167–175° F [75–79° C], similar to common resin but more expensive; makes a smoother pouring mixture.
6. Petroleum jelly, used to lower M.P. and helps eliminate lap lines in pouring mixtures.
7. Mineral sprits, lowers and helps render wax mixtures more homogeneous. Be careful when adding to melted mixtures; **remove wax pot from the stove before adding.**

Modeling Wax

Microcrystalline wax usually comes in 10-pound [4.5 kilo] slabs. To break the wax slab into smaller pieces, score a line with a sharp knife on three sides of the slab where you want to break it; by striking the slab against the edge of a table, it should fracture right at the score line. At room temperature, the wax is too stiff to model by hand. It can be softened by putting small pieces into a bucket of warm water, using a heat lamp, or setting the wax in the sun, melting the wax and then allowing it to cool until it is easily moldable. Warm, soft wax should have the consistency of soft clay and be easily pushed around by your fingers. Wax modeling is done in a similar manner to modeling clay. When adding one piece to another, care should be taken to achieve a good join: the two pieces should be at the same temperature and no dirt or water should be between the two parts. The wax can also be manipulated with knives, dental tools, or other instruments. The sculpture can be kept warm by returning it to a warm water bucket. Be sure to blow off the water before resuming the modeling process. To keep the wax from distorting during periods of warm temperature, float the wax in a bucket of cool water or store in a refrigerator.

Different textures on the wax surface can be achieved by a variety of methods. Scraping the wax with a serrated modeling tool or grapefruit knife is a good way to begin smoothing and clarifying the form following an initial hand modeling. Melting the surface with a candle, propane torch, hot knife, electric soldering iron, dipping the wax in a container of melted wax, or brushing with melted wax will all make for a variety of surface effects. Wax can also be smoothed by sanding the cool wax with sandpaper or used pantyhose material. Use a small amount of mineral spirits to facilitate the sanding process. Wear gloves when working with mineral spirits.

Traditional modelers in Africa and India use thin wax threads modeled over clay cores to achieve their unique results and control thickness.

Casting Wax in Molds

By using molds, the sculptor can not only make a wax pattern from a sculptural form executed in another material, but multiple images can also be made. Plaster molds are the most frequently used molds for most sculptors.

The cup mold and the flat plaster "batt" are two examples of "open" or one-sided plaster molds. Two piece (or multiple piece) plaster molds can also be used, provided they have an opening sufficiently large in which to pour the wax. With a plaster mold, the mold must be thoroughly wet before pouring wax into the mold. Wax will stick to dry or inadequately soaked plaster molds.

Using the plaster batt to make sheets: Soak the batt in the sink for at least 30 minutes if it is completely dry. The batt should feel damp and cold when held to the cheek, or put the batt to your ear, if you hear faint "sucking" noises, the batt is still absorbing water. The batt should be placed so that it is level. Blow the excess water out of the bat with the air gun or blot with a cloth. Melt the wax in your wax pan or electric wax pot. For best results use a meat or candy thermometer to determine the temperature of the wax. Do not allow the wax to exceed 200° F [93° C]. Very hot wax will cook the water out of the batt and scorch the mold surface. If the mold turns light brown on the surface, you have burned its surface—discard the mold. When the wax is about 170° F [77° C], it will be clear and shiny on the top surface. The wax will turn cloudy and dull on the surface as it cools down to its melting point. If you are not using a thermometer, pour the wax in the batt just as the wax begins to cloud up on the surface. Textural effects other than smooth sheets can be created by pouring very cold wax on the batt or by pouring a small amount, stopping, and then re-pouring or by pouring wax into water.

Slush Molding Plaster Molds

The cup mold is a good example of a one-piece mold which is easily slush molded. Soak the cup mold in the same manner as the batt. Remove excess water and fill the mold with wax that is just above the melting point. Observe the wax at the sides of the mold. The wet and cooled mold will begin to chill the wax at the mold interface. When the wax has chilled to a thickness of 3/16"–1/4" [.6 cm.], carefully pick up the mold and dump the remaining hot wax back into the wax pan. Immersing the mold in a bucket of cold water can quickly chill the remaining wax casting. The wax will shrink slightly, enabling you to easily remove the wax piece.

The same method can be used to cast hollow waxes in two-piece or multiple piece plaster molds. Again, soak the mold pieces. Multiple piece plaster molds should be held together with stout rubber bands made from inner tube rubber. The rubber bands not only keep the mold from warping when it is not being used, but hold the pieces in their registered positions. Make sure than the opening into the mold is 4 times the wall thickness of the wax you want to cast. Support the mold securely and carefully pour the wax at the correct temperature in to the mold. Wait until the wax has chilled to the desired thickness and pour out the remaining wax. "Lap lines" will be formed on the wax surface if the wax is too cold or the pour rate is slow. To cast large molds without having to melt large amounts of wax, quickly pour the mold about 1/3 full and then rotate it, causing the wax to coat the inner sides of the mold. Continue rotating until the wax has chilled to the correct thickness.

If the plaster mold has delicate detail or has parts of the surface that stick up into the mold cavity (deep depressions in the casting), it may be easier to paint the wax over the detail and the raised sections before closing the mold. Some sculptors use two different batches of wax at two different temperatures to get intricate detail out of a mold. The first painted-on coat of wax contains petroleum jelly and is slightly softer to help eliminate lap lines while applying the wax; this wax is also kept hot enough to eliminate the lap lines. The back-up wax is not only cooler, but may contain paraffin to stiffen the wax casting. Raised sections in the mold cavity (especially if they are thin) will be difficult to coat as they heat up quickly causing the wax not to cool over these parts, creating thin spots. After the initial coat of wax to capture all the detail has been applied, close the mold sections (avoid getting wax on the parting surfaces or the mold will not go back together evenly) and slush cast the cooler, stiffer wax. Check for thin spots after the wax casting has cooled by holding the wax up to the light; the thin spots will be seen as light brown areas as the light passes through the thin wax. Usually it takes a number of tries with both casting methods to get a good wax pattern.

Casting in Latex Molds

More complicated sculptures can be molded with latex rubber using an "envelope" mold (the mold has one seam in the latex on one side of the sculpture). The latex envelope mold eliminates the many parting lines that would have to be chased off of a wax pattern cast in a multiple-piece plaster mold. Latex molds are thin rubber backed by a thin (1" or less or 2.5 cm. or less) "mother mold" which holds the latex in the correct form. Consult the *Bibliography* for books containing information on general sculptural practice, or the latex manufacturer's instructional material.

Soak the mother mold so that wax does not stick to it during the slush molding process. Open the latex mold and spray the inside surface with a light coat of silicon spray. Reassemble the mold, making certain the latex is in its proper place, and band the mother mold sections together. Since the latex heats up faster than wet plaster, be sure that the wax is at the lowest possible temperature to achieve the proper detail. Pour the wax in and coat the inside surface of the latex mold with a thin layer of wax and then pour out the remaining wax. Allow the wax to cool slightly and then pour cooler wax (cloudy surface or just above the melting point) into the mold and slush again. Continue slushing until the correct thickness of wax has been built. Cool the wax by immersing the entire mold in cool water. Remove the mother mold and then strip the latex from the wax casting by peeling back the latex from each side of the seam. The wax will have a coating of silicon, which should be removed or it will interfere with additional modeling and the spruing and gating process. The silicon can be removed by gently washing the wax with soapy water

Casting in Silicon Rubber Molds

Silicon rubber molds are appropriate whenever many copies of the wax pattern is desired. Silicon rubber is significantly more expensive than latex and other rubber molding materials. The advantage of silicon rubber is its resistance to heat, the ability to get thousands of casts without breakdown of the rubber, and the fact that wax does not stick to it, so that no separator is necessary. Consult the manufacturer's information concerning the compatibility of the rubber with various pattern materials. Poly-Tek Development Corporation [http://www.polytek.com/] is good source for various rubber molding materials, plus its website has numerous tutorials on making rubber molds and excellent technical information.

Silicon rubbers are both flexible enough to be able be stripped easily from both pattern and casting and are stiff enough to be used without a mother mold. Most silicon rubbers are pourable, but some can be buttered on. For a very cheap silicon rubber mold, use the silicon rubber caulking compound sold in a caulking tube and back with a plaster mother mold. See the *Appendix A* for information on using silicon caulk for molding.

Other Types of Rubber Molds

For rubber molds which are more expensive than latex but cheaper than silicon rubber, try rubber made with polysulphide compounds, such as "Black-Tufy" made by the Perma Fiex Mold Company, 1919 East Livingston Avenue, Columbus, OH 43209. [http://www.perma-flex.com/]. Black Tufy compound is a polysulphide rubber that contains no water, so it will not dry out or shrink, thus retaining its dimensional accuracy for long periods. It is not affected by water, oil, or most solvents. Black-Tufy is mixed with a two-part curative agent and must be thoroughly mixed. Read the manufacturers directions thoroughly and consult the instructor for advice as to how to set up the pattern for molding.

For these types of pourable rubbers, the pattern is firmly anchored to a bottom board so that it will not move out of position. It is covered first with a clay that determines the thickness of the rubber, usually about 1/4–1/25 [.6–1.25 cm.] and then a seam ridge of clay and a clay pouring sprue are attached. A two-piece plaster mother mold is then made. After the plaster is set, the mother mold is removed and the clay blanket is stripped away from the pattern. The mother mold is reset in position around the pattern, sealed, and then the rubber poured in. Following the cure time, the mother mold and cured rubber mold are stripped from the pattern, cleaned, and reassembled. The mold are now ready to be poured with wax to make the wax pattern.

Polytek Development Corporation, P.O. Box 384, Lebanon, NJ 08833 (908) 534-5990, [http://www.polytek.com/], makes a nice rubber that can be buttered on to the thickness desired. "Polygel" is available in two consistencies, Polygel40 and Polygel50. Both RTV (Room Temperature Vulcanizing) rubbers come in two-part liquid components, which when mixed together, change color and thicken to a buttery, non-sag paste. The first coat is buttered on to a thickness of about 1/16" [.15 cm.] and allowed to cure for one hour. The second coat is then added until the thickness is between 1/8"–3/16" [.3 to .45 cm.]. Allow to cure for 24 hours, then make a mother mold of plaster. Polygel molds are tough with a high-tear strength, resist grease and oil, are dimensionally stable, and have a long storage life. Consult the manufacturer's directions for sealing patterns and mixing directions.

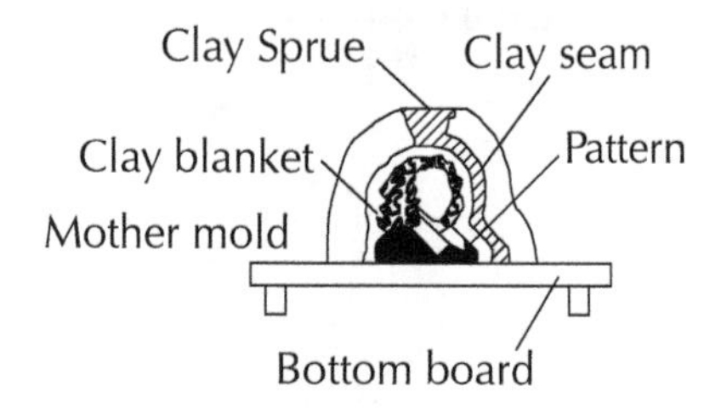

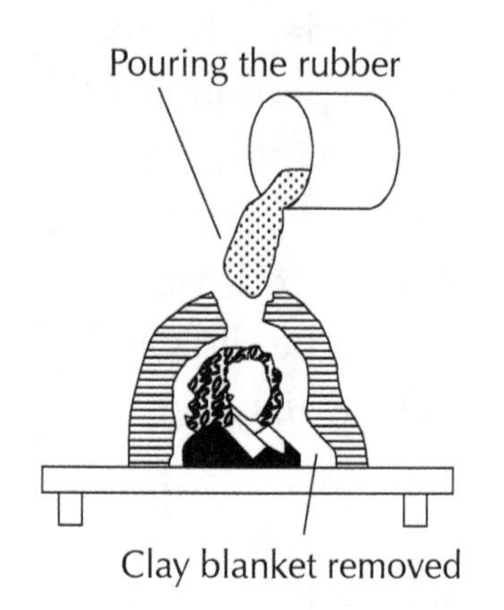

Casting in Clay

Wax will not stick to wet clay, making it a fast and inexpensive mold material. Its ability to capture textural detail and simple form means that textures and simple forms can be cast into wax by pressing objects and textures into a slab of clay. The clay should not be too wet or too dry, but at an easily moldable consistency. Pour or brush wax into the depressions and forms created. Allow the wax to cool and remove the wax castings. Any remaining clay can be washed off of the wax.

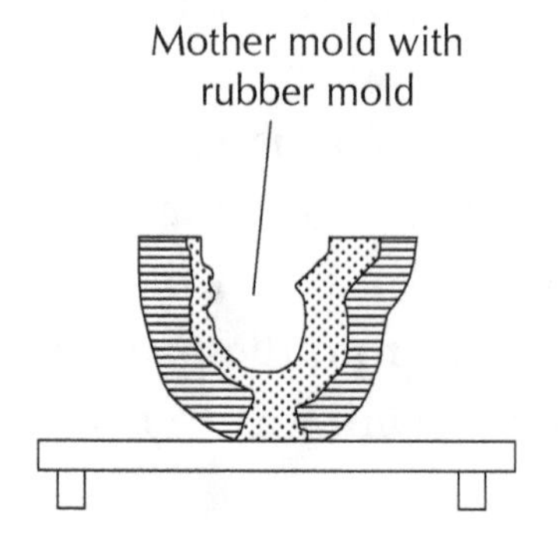

Construction with Wax

Wax sheets, modeled forms, and cast forms can be joined together by welding the wax. Thin wide knives or spatulas work the best for welding. To weld two pieces together, heat the knife over a flame (stove or propane torch) until it is hot enough to easily melt the wax. Place the two cool pieces that are to be joined together in the correct position, separate them slightly, and inset the hot knife. Push the two pieces together while slipping the knife out of the join. The wax should have melted on both pieces. Hold in position until the wax has cooled. It is extremely important to learn to weld wax, as wax sprues, cups, and vents must all be joined on to the wax pattern through welding.

The use of a very sticky wax will aid in assembling tentative compositions without having to weld parts together. To make a "tack" wax, add petroleum jelly and turpentine to the microcrystalline wax until it is soft and sticky when thoroughly cooled. Test the wax by cooling a small amount and adjust additives to achieve the consistency desired.

Welds can be strengthened by using a hot soldering iron (old style or electric) to melt the wax along a seam. Very long welds are difficult to do and must be reinforced. Brushing hot wax on a seam is another way to strengthen it. A good tool for welding can be made by brazing a thin blade on an electric soldering iron; the knife stays hot without having to return it to the flame.

Expanded Polysyrene Patterns

Expanded Polystyrene (EPS), also called styrofoam, can be used as a pattern material in both the lost foam process or as a pattern for self-set sand. EPS comes in sheets (used as an insulating material in building construction), in a rope-like form of varying diameters (used as a caulking material), as small forms such as "peanuts," "shells," or "beads" (used as a packing material), in various simple forms like cones, balls, or bells (found in hobby and craft stores), and in special forms which are usually packing supports. EPS can be both foamed in structure or composed of expanded polystyrene beads. If using the EPS in the lost foam process, select the foam which has the lowest possible density (1.0 pound/cu. ft.), and be sure that the foam is not the blue-colored fire retardant variety or the light brown-colored urethane (both of these foams will not bum out). Both the fire retardant and urethane foams can be used as patterns with self-set sand.

EPS and urethane foams can be easily cut with a band saw, knife, hand saw, or hot wire. Special glues should be used if you are using the lost foam method of casting, see the section "Evaporative Foam Casting Process" in chapter 7. For patterns to be used in self-set sand, white glue or rubber cement will work. Use straight pins to hold pieces together while the glue sets. Textural detail can be achieved on foam with hot tools, spray enamel (which dissolves the foam), "magic markers" (which also dissolves the foam), wax application; paper, plastic wrap, or aluminum foil can be glued on the surface.

Plasticene Patterns

Plasticene is clay that uses oil, grease, and wax as its binder rather than water. It does not dry out and is readily reusable. Plasticene is useful when using the self set sand method as a liner to determine metal thickness and also as a way to false piece small patterns. Self-set sand molds can be made directly off of modeled plasticene patterns. Commercially made plasticene is an expensive material, but it can be easily made in the studio. To make plasticene, use the following recipe:

20 pounds plasticene [9 kilos]

Microcrystalline wax..... 5 pounds [2.26 kilos]
30 W mineral oil............ 2.5 pounds [1.13 kilos]
Grease............................ .5 pounds [.226 kilos]
Clay (Air float bond)....... 12 pounds [5.44 kilos]

Use a 5 gallon [19 liters] metal pail or a large metal pot to make the plasticene. First melt the wax and then add the oil and grease into the liquid wax. Stir the mixture to dissolve the grease. Add the clay by sifting small amounts into the wax, oil, and grease mixture (Wear your dust mask!). Do not dump clumps of clay into the mixture, as lumps will result. Continue to add the clay and stir until the mixture is smooth. Heat until the mixture begins to boil and bubble. Remove a small portion and cool. Test the plasticene for its consistency. If it is too stiff, add more grease. If too greasy, add more clay. Adjust the mixture to suit your needs. Pour the mixture into a cake pan or on to a metal sheet lined with aluminum foil and cool.

The plasticene should have the same consistency as well-wedged water clay. Plasticene can be used by itself or molded over other materials to form the pattern. When using plasticene as a pattern with self-set sand, part well with graphite or talc before ramming the sand and do not allow the pattern to remain in the sand for extended periods of time, as the oil of the plasticene will react with any resin-bonded sand and weaken the sand mold.

Wooden Patterns

Wooden patterns were the traditional patterns used in commercial metal casting. Pattern makers were the "kings" of metal casting as their specialized skill in translating drawings into accurate three-dimensional patterns made them indispensable. Today with the development and use of digital 3-D printers and other new plastic materials, wooden patterns may become a practice of the past. Most wooden patterns are mounted on pattern boards for use in green sand molding, although "loose" or un-mounted patterns are also used. Depending on your concept, wood patterns may be the way to easily realize your vision, especially if you intend to reproduce the pattern more than once.

Traditionally such fine woods as mahogany or #1 clear pine have been used as pattern materials. If you are using a lathe to turn the pattern, #1 clear pine is especially recommended. However, regular pine and plywood can be used, depending upon the idea you are attempting to realize. Patterns made from wood are classified as rigid patterns and must have "draft." Draft is defined as a taper of the sides of the pattern that allows it to be easily withdrawn from the sand mold. To determine the weight of the casting to be made from the pattern, weigh the pattern and then consult *Appendix B,* "Weight of Pattern to Weight of Casting."

Wood patterns should be painted with lacquer and rubbed with graphite. Red lacquer is the traditional color for patterns. Do not paint the patterns with enamel as the resin in the binder will make the sand stick to the pattern.

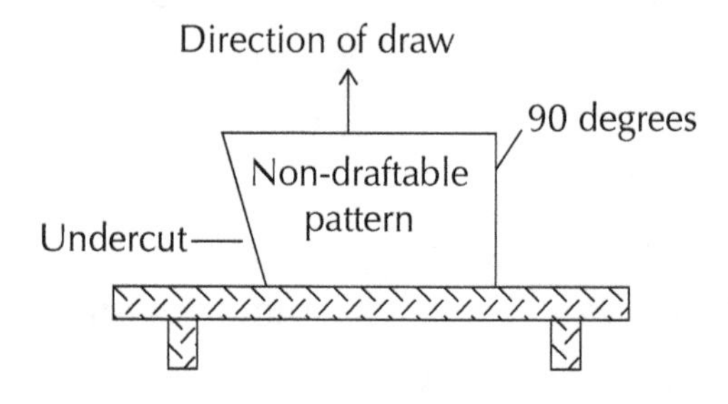

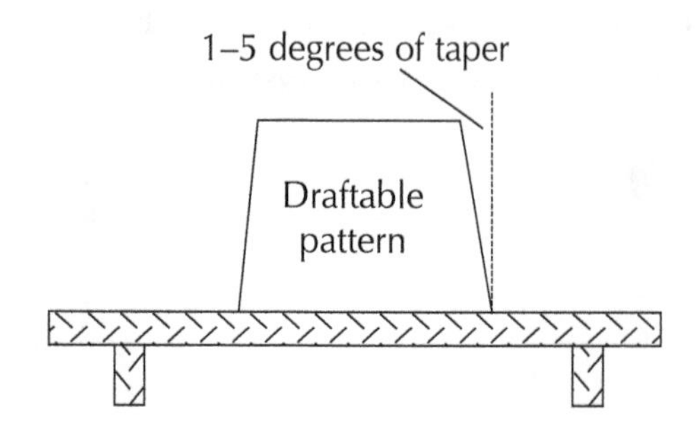

Split patterns are loose patterns that are divided along the center-line of the pattern. One-half of the pattern is loosely fixed to a follow-board to make the first half of the mold. The mold is rolled over and the follow-board is removed. The other half of the pattern is registered (usually with dowel pins), on the first half of the pattern already embedded in the sand. The second half of the mold is then completed.

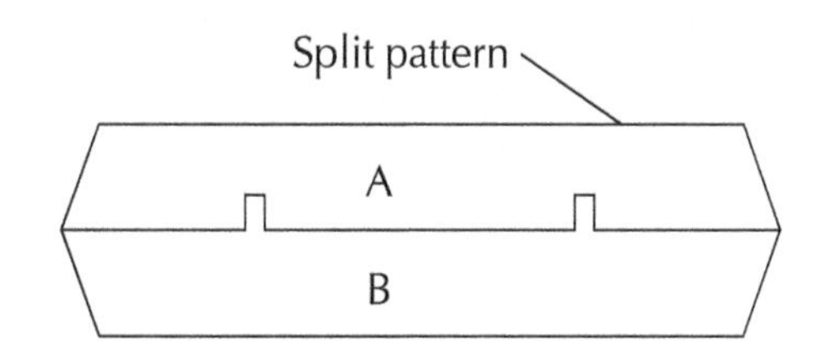

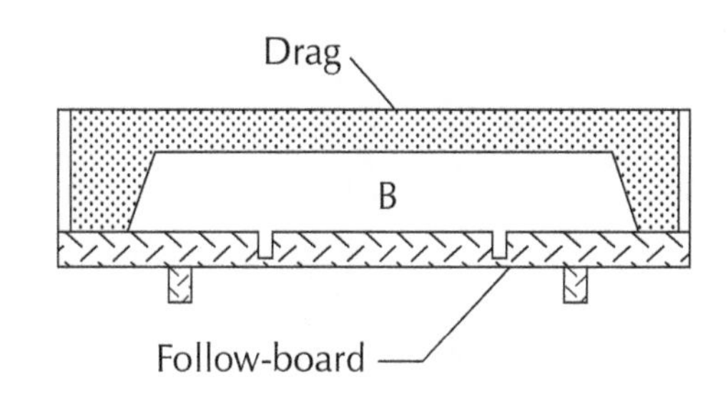

Other Pattern Materials

Almost any material can be molded. Other rigid materials such as metal or rigid plastic should be false pieced to get the first mold off of the pattern. Some materials such as glass do not release very well from self-set sand. These patterns should be sprayed with a mold release such as graphite and alcohol to facilitate release from the mold. Consult the section "Self-set Sand" in chapter 7, for a discussion of false piecing. Flexible materials such as soft plastic, rubber, cloth, rope, cardboard, paper, or even leaves, plant materials, fish or other organic materials can be molded in self-set sand, as these materials can easily be pulled out of the sand. Part these materials with graphite before molding. Only the limit of your imagination and vision restricts the kind of pattern materials you can employ.

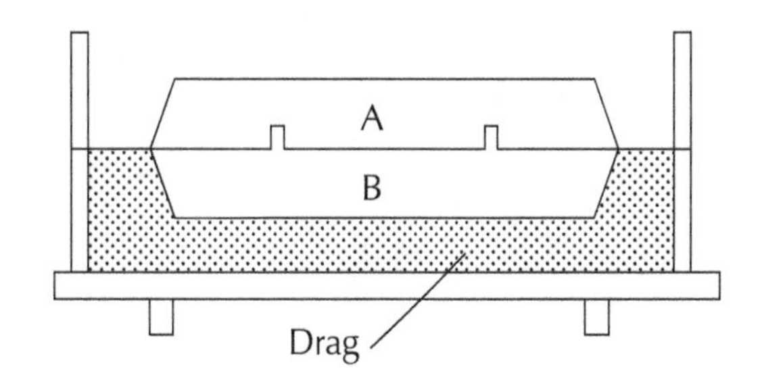

Spruing and Venting Wax Melt-outs or Burn-outs

Chapter 6

A "sprue" is a main channel that delivers metal from the pouring cup to the casting. The "gate" is the point at which the metal enters the pattern cavity of the form that is to be cast. A "runner" is a secondary channel that also delivers metal to the casting. A "vent" is a channel on the top of the casting which allows gasses generated in the casting process to escape the mold. A "riser" is a thick vent that also feeds metal back into the casting as the metal shrinks during cooling; risers are usually positioned over thick parts of a casting.

Commercial wax sprues can be purchased which are round, hollow, and usually red in color. However, it is more economical to make your own wax sprues, runners, and vents. Most sprues should be between 3/4"–1" [1.9–2.5 cms.] square. To make your own sprues and vents, fill a deep plaster batt with hot wax to a depth of 3/4" or 1". Allow the wax to cool and then remove the thick sheet. Cut the sprues to the correct thickness; vents can usually be thinner than sprues, about 1/4"– 1/2" [.6–1.3 cms.]. Hollow sprues can be made by piece molding a round or square dowel in plaster and then by slush molding the wax in the plaster mold

Because most sculptures are unique, there are few hard and fast rules about spruing and gating your piece. Designers of commercial cast objects often employ computer programs that model the flow and cooling of the metal in the mold to determine correct spruing and venting of the casting. This is probably beyond the capability of many sculptors; experience and familiarity with the metals you are casting will help determine the correct spruing and venting system.

Remember: Every sprue, runner, and vent which is attached to the wax pattern must be cut off when the casting is transformed into metal.

There are two main ways to sprue and vent a wax pattern: the bottom pour system and the top pour system:

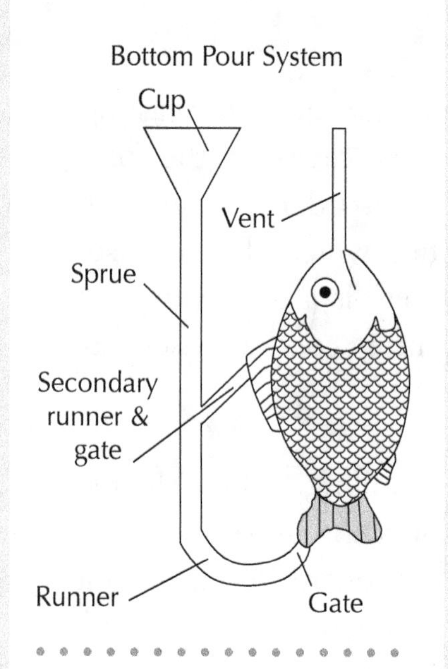

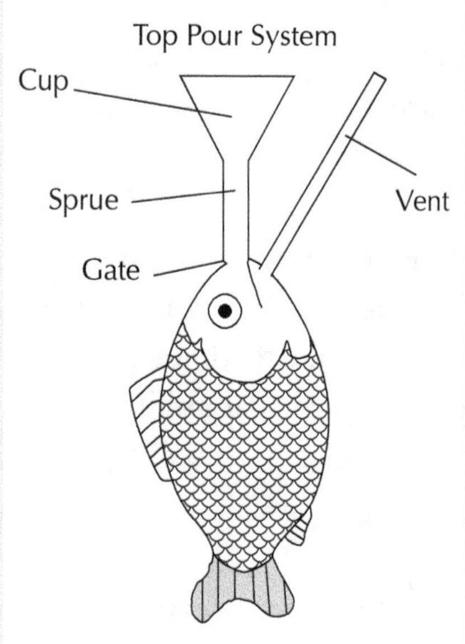

Avoid attaching sprues and vents to detailed sections of the pattern or in depressions in the pattern; the metal will be easier to chase off if the sprues and vents are at the high points and at places of little or smooth texture. Core openings or "necks" should be facing up in the mold to facilitate the escape of gasses in the core. Thick sections of a pattern should have risers or thicker vents. In general sprue systems are "tree-like"—the cup is the thickest, the down sprue is thinner, the runners thinner yet, and gates are about the thickness of the pattern thickness. Avoid radical thick-to-thin sections in the pattern or in the sprue system. Although there are mathematical models that can determine all of the dimensions and cross-sectional areas of sprues, runner, and vents, the uniqueness of most sculptural patterns means that most sculptors rely on their experience, intuition, and luck!

In the "bottom pour system," metal enters the mold cavity at the bottom and fills the cavity as the level of the metal rises from the pressure on the cup. Secondary runners and gates feed extremities or difficult sections and take over as the level of the metal rises to their height. Note that secondary runners slant up to prevent the metal from entering the casting initially. Vents are at the top, allowing gases to escape. In the "top pour system," the metal enters the mold cavity at the top of the pattern and fills it up. Vents are located at the top of the cavity near the sprue. In both systems, making the gates slightly smaller ("choking the gate") increases the pressure during the pouring.

The choice to use either the basic top pour or bottom pour system is based on the form of the pattern you have made and the kind of mold material being used. When casting with ceramic shell, the top pour system is almost always employed and no vents are used because the gases are vented directly through the shell. The advantage of the bottom pour is that the metal does not wash over the detail of the mold interface as it fills the casting, but rather the metal rises evenly throughout the mold cavity because of the pressure on the system from the full cup. The advantage of the top pour is that there are usually less gates to clip off and chase afterwards. Some patterns may need more than one down sprue. These systems are called the split bottom pour and the split top pour system

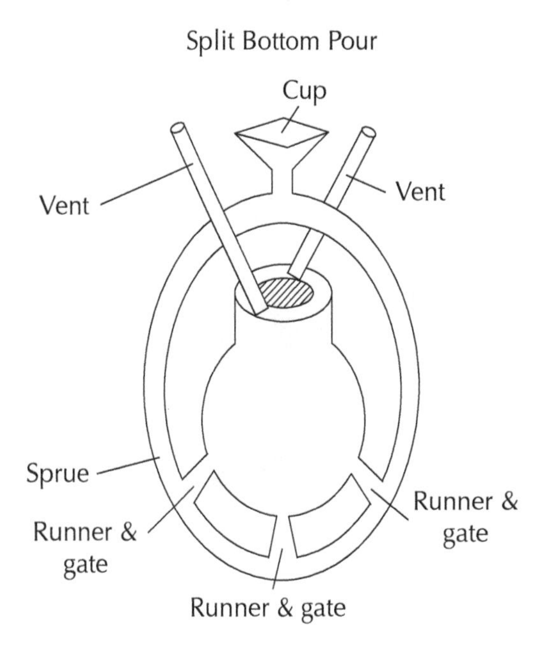

When attaching the square wax sprues and runners, be sure that they are smooth and that the sharp edges have been scraped down. Everything about the system should be slightly curved and smooth. Fill in the sharp angle where the sprue joins the runner with a fillet of wax. Sharp angles in the sprue system not only create turbulence, but they make for sharp mold sections that can heat up and break free during the pouring, causing inclusions in the casting. Be sure that everything is securely welded together; you should be able to pick up your piece by the sprue system when the wax is cool. Nothing is more disheartening than to have parts of the sprue system fall off as molding begins!

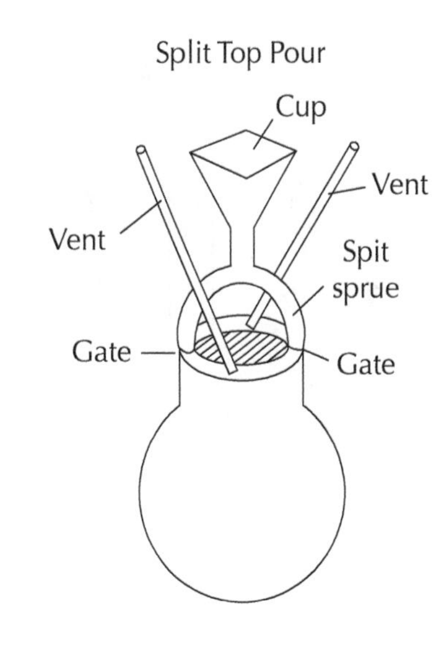

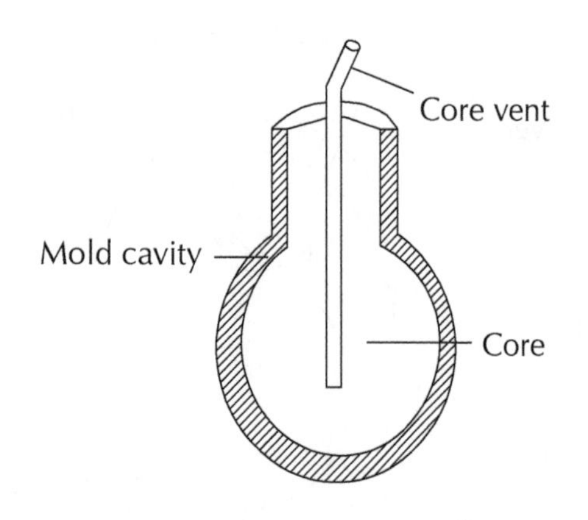

Large cores might require core vents that lead from the interior of the core to the top of the mold. A 3/8" [.96 cms.] wax rod or a piece of flexible copper tubing inserted in the core will allow core gases to escape through the vent. The core vent should go all the way to the mold exterior and not touch the sprue system or the mold cavity, or it will fill with metal and not function to remove the gases.

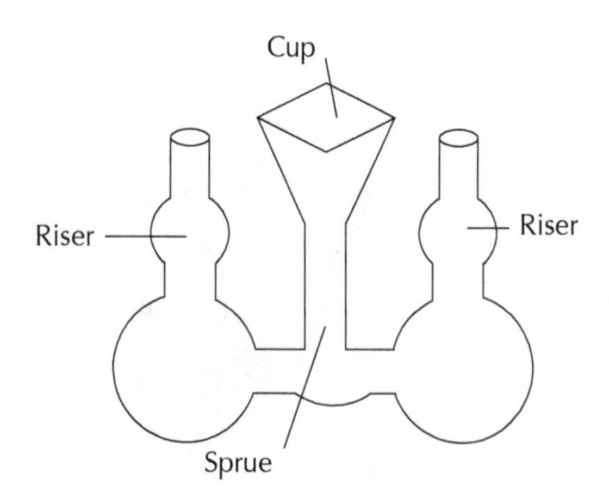

Risers are specialized vents which function not only to allow the gases to escape from the mold, but prevent cracks and distortion as the metal shrinks during cooling. Risers are usually placed over thick sections, especially when these sections are next to or near thinner sections. In castings with irregular thickness, the thicker parts remain molten longer and shrink more. The thick riser holds a reservoir of hot metal which feeds back into the casting as the casting shrinks. Commercially available exothermic risers are useful for very thick sections of a casting, these risers keep the metal in the riser hotter for a longer period of time.

Spruing for Ceramic Shell

Because the ceramic shell material is porous, gases generated during the casting process are vented directly through the shell, so vents are usually not required. A simple top pour system that is as direct as possible is usually the best way to sprue a wax pattern for ceramic shell. Use a short sprue about 2–3" [5–7.6 cms.] long and about 1" [2.5 cms.] square. Securely weld the sprue to a thick area of the pattern that can be easily chased after casting. Using a hollow sprue can help facilitate the attachment of a hollow cup to the sprue. A hollow cup allows the wax in the sprue and pattern to flush from the mold more quickly during the flash-out process, eliminating possible cracking from rapid wax expansion as the wax goes from solid to liquid. Using a Styrofoam cup and a hanger made with a lag screw on the end works well. Fill the outside bottom of a Styrofoam cup with wax and allow to cool. Weld the cup to the sprue and allow it to cool. Then dip the lag screw end of the hanger in hot wax and screw down through the Styrofoam cup and securely into the hollow sprue. Add a bit of liquid wax to the inside bottom of the cup. Alternatively you could make a solid wax cup with an S-shaped hanger made from a 1/4" [.6 cm.] rod embedded securely in the wax. You can facilitate the elimination of the wax from a solid cup by using a torch to melt out most of the wax from the completed shell before putting it in the flash-out kiln.

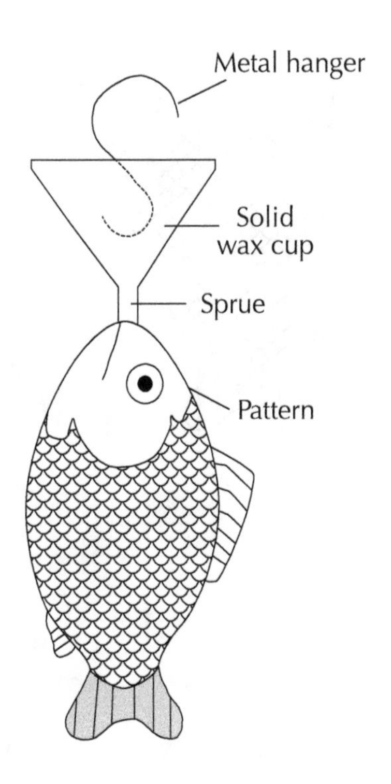

Thin finger areas or other extremities may need to be tied together with thin wax rods to make sure that the ends fill out. If spruing many small parts to one sprue (gang spruing), make sure the pieces are pointing downward. Parts that unavoidedly must be pointing up should be vented to the top edge of the cup.

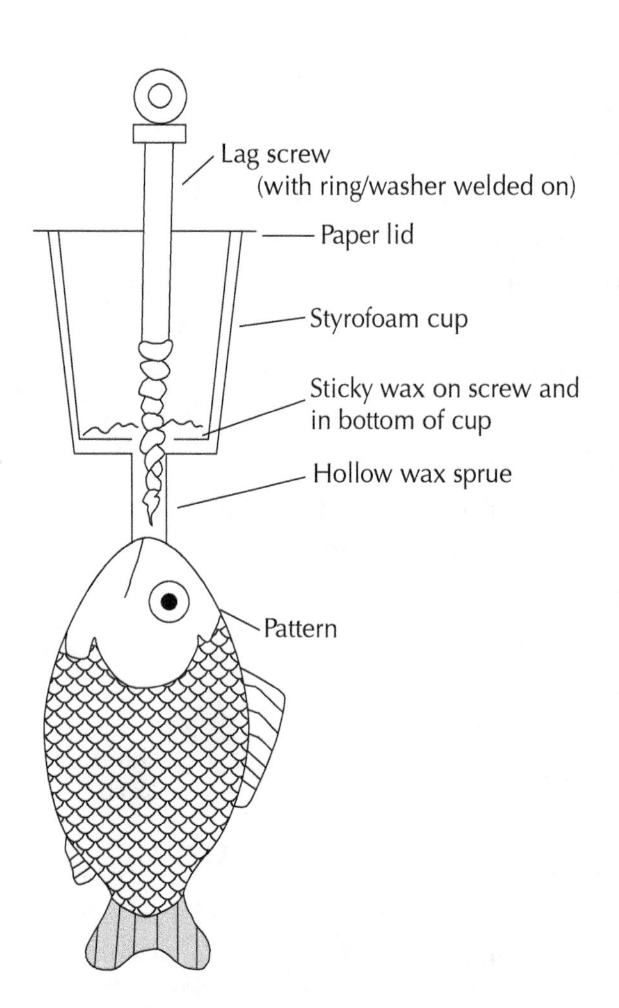

Spruing and Venting Sand Piece Molds

In most sand molds, the sprue is cut through the cope (top) half of the mold and terminates in a well in the drag (lower) half of the mold. The well is connected to runners that can either be in the cope or in the drag (depending on the metal being poured). The gates are cut in the cope and overlap the runner and the mold cavity. A sprue system can be conceived of as a tree: the sprue is the trunk, the runner is the main branch, and the gates are the smaller branches. The system steps down from big cup to thick sprue to thinner runners to thin gates. The gates are usually the same thickness as the thickness of the casting (sometimes thinner).

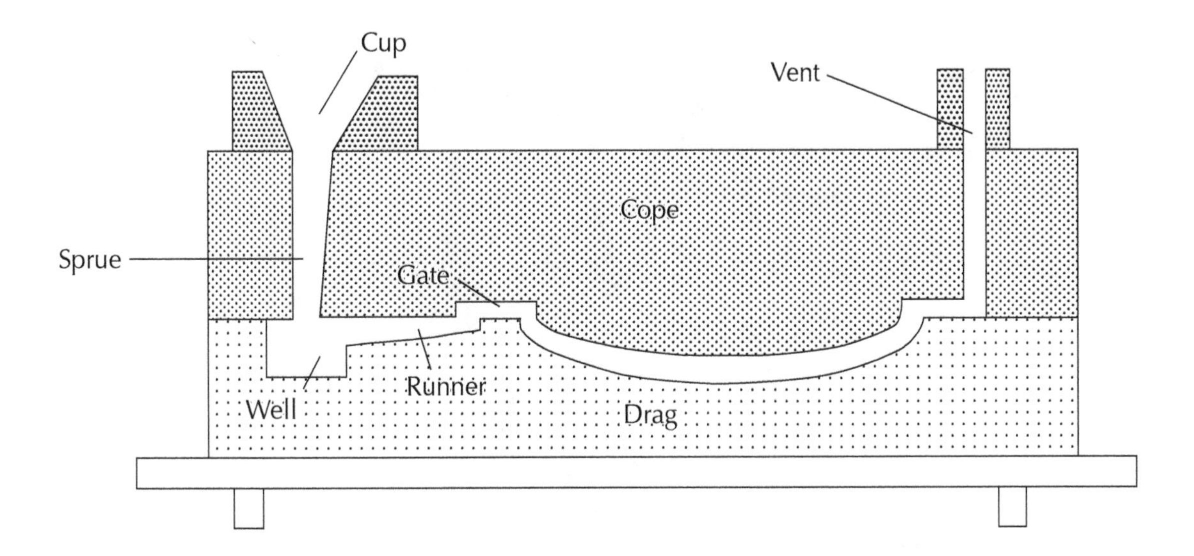

In general, the following guidelines will help you determine the size of each of the parts of the system:

A. The base of the sprue should be about 1/3–1/2 of the cross-sectional area of its top.

B. The well is approximately 5 times the cross-sectional area of the sprue base and 2 times the depth of the runner.

C. Runners should be wide and shallow; approximately 3 times as wide as deep. An extension on the runner past the last gate keeps the first metal (which is always cold and dirty) from entering the mold cavity.

D. Gates should have about the same cross-sectional area as the base of the sprue, but should be wide and thin. The first gate should be within 6" of the well.

E. The total cross-sectional area of the runners and gates should be approximately 2 times the area of the sprue base.

Some patterns in self-set sand can be top poured. Bells are traditionally cast in this manner.

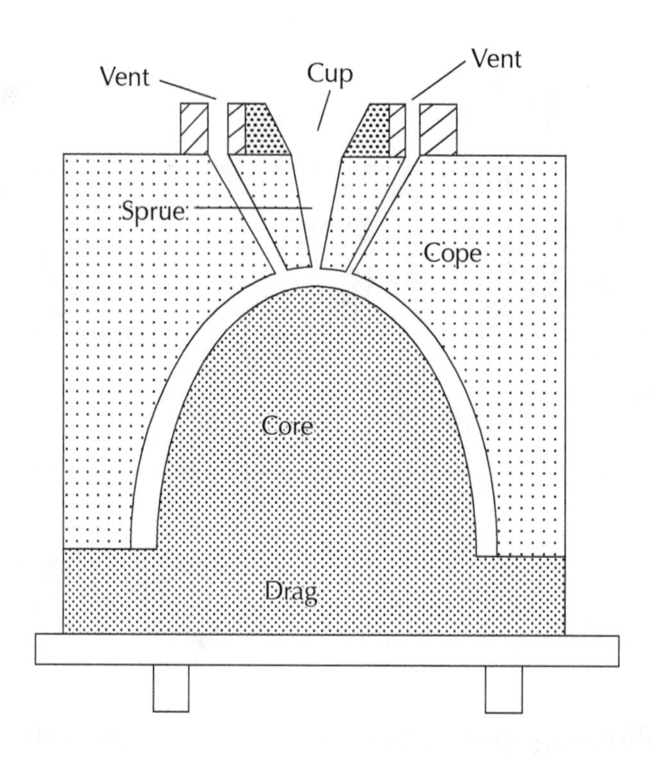

With green sand and Petro Bond molds the sprue can be cut with a piece of thin-wall metal tubing about 3/4"–1" [1.9–2.5 cm.] in diameter. Whistler vents can be made with a 1/8" [.3 cm.] metal rod pushed through the sand. Runners and gates can be cut with a thin piece of sheet metal bent in a U-shape. Wells can be scooped out with a spoon. Self-set sand is considerably harder than green sand. Use masonry bits in an electric drill to drill vents and to make pilot holes for the sprue. Wide wheel-shaped mounted carborundum stones are useful in cutting the well. Use cone-shaped mounted carborundum stones with a 1" diameter at the base to finish off the sprue. The cone shape is handy because it makes a natural choke at the bottom of the sprue. When molding with self-set, be sure to make the flask big enough to be able to cut the runners and gates. Remember: there should be at least 2" of wall thickness between the metal and the outside of the mold and more space if the piece is very large. Cups and vents must be the same height to maintain pressure on the casting and to prevent metal from running out of the mold and on to the floor.

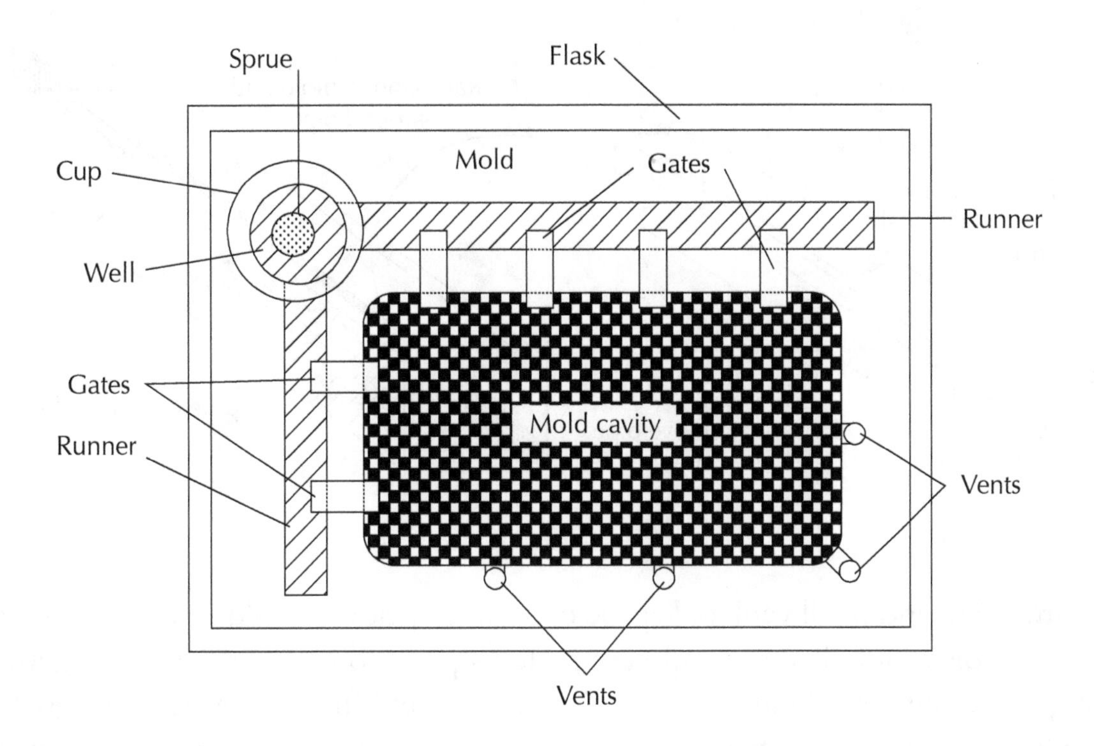

Cups can be made in the cup molds; sand sprues and vents are made with a sprue bar mold. Be sure to remove the pipe from the mold after ramming the sand. as it cannot be removed once the sand is set! Vents and sprues can also be made by drilling through rectilinear pieces of self-set sand.

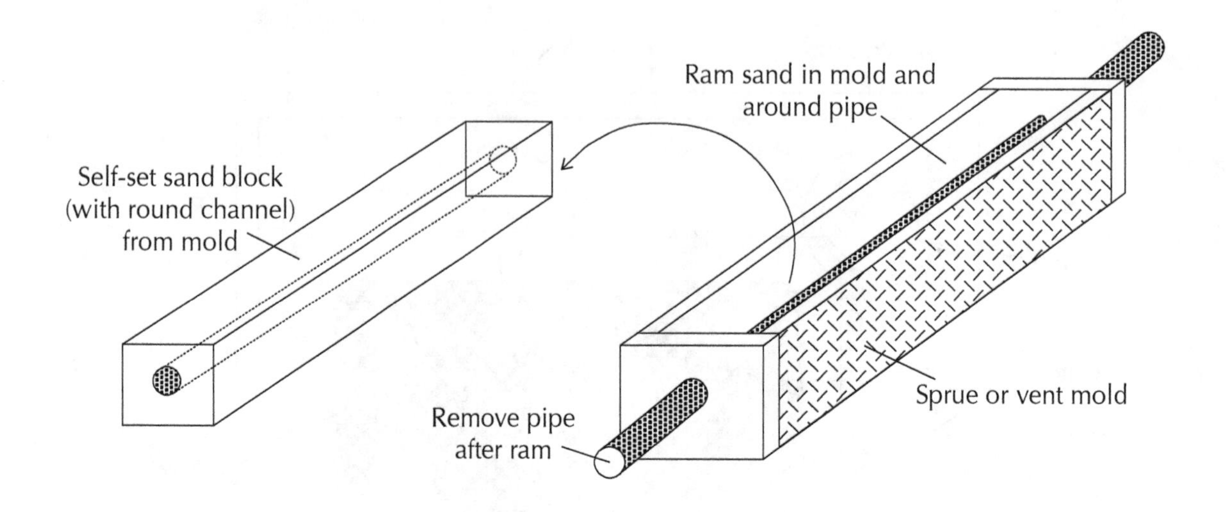

Be sure to cut and fit all vent and sprue extensions prior to mold assembly. Use an air gun to blow out all sprue and vent systems prior to gluing the mold together. Use good quality mold glue such as "Green Hornet" [http://www.hillandgriffith.com/green-sand/]. Construction adhesives used in a caulking gun will also work. Glue paper over the cup and vents to prevent dirt from entering the mold cavity during handling and placement on the floor. All molds should be identified with your name, type of metal being poured, the estimated weight of the casting including sprue, well, runners, gates, and vents], and whether the pattern cavity is "thick", "medium", or "thin".

Introduction to Molding

Chapter 7

Good mold making is the key to a successful transformation of your pattern into metal. All molding systems employ some sort of refractory, such as sand, which resists the heat of the metal and some sort of binder to hold the refractory together and form the mold around the pattern. In the following sections, you will be introduced to a variety of the molding systems used by sculptors. The type of molding system used is a function of the type of pattern you want to cast, as some systems are better than others in certain pattern situations.

Remember that all molds are enantiamorphs, that is they are three-dimensional mirror opposites of the positive object. What is on the right in the mold will be on the left in the casting and what is up in the mold will be down in the casting.

"No-bake" or "Self-set" Molding Processes

Furan- These systems use furfuryl alcohol-based binders catalyzed with acids to produce a tough resin at room temperature. Furan is a two-part system: a binder and a catalyst. The advantage of furan is that it produces strong molds, has a short strip time (strip time is the amount of time before the pattern can be with drawn from the mold, prior to the time needed for the mold to fully cure), curing time, changes color when cured, and shakes out easily. Its disadvantages are that is sticky to ram, produces an odor during curing and pouring, is caustic, and difficult to clean up.

Alkyd oil- These systems use oils that are catalyzed with isocyanate to also form rigid resin at room temperature. These binders are either two-part or three-part systems. In the three-part system an accelerator is added which controls strip time and cure rate. In the two-part system, the accelerator is pre-added to the binder.

The advantages of alkyd oil systems is that they produce less odor than furan, have faster strip times, and form a flexible bond with the sand facilitating an easier draw of the pattern. The disadvantages are that it is not as strong as furan, is also caustic, is smoky during pouring, and is harder to shake-out. There are many "no-bake" or "self-set" systems available. HA International [http://www.ha-international.com/content/products/resins/nobake.aspx] or United Erie [http://www.unitederie.com/] have a variety of no-bake resins.

Sodium silicate- In this system, liquid sodium silicate (also called "water glass") is used to bind the sand grains together. Organic materials such as wood flour and sugar are sometimes added. This binder is set up by either gassing with C02 or using an acid catalyst. The advantage of sodium silicate is that has a low toxicity, cleans up with water, is cheaper to use, and produces little smoke during pouring. Its disadvantages are that has a low tensile strength, is friable (the tendency to be brittle and crushable), the molds cannot be stored for long periods as they pick up moisture which breaks down the sand bond, and it is difficult to shake-out. Sil Bond (catalytic set) made by United Erie is a frequently used product.

Mixing the "No-Bake" Resins

The sand should be dry and cold as wet sand will resist the binder—producing a poor bond—and heat accelerates the bench life (working time) of the sand mix. Grain size of the sand used will determine texture and detail. The larger the grain size, the more resin must be used. 30-40 mesh silica sand works well with most applications. Patterns with fine detail will require finer sand (50-80 mesh).

Remember, to wear a dust mask during all digging, handling, and mixing operations! Turn on the ventilation systems before digging sand.

A molding bench as well as rammers, strike-off bars, wedges, riddles, and sprue-bar molds will facilitate most sand molding work. Disposable gloves are recommended during all mixing and handling operations. Recycled 1/2 gallon [2 liter] milk cartons can be used to weigh the binder; recycled juice containers are useful to weigh the catalyst.

The amount of binder is dependent upon the type of metal being poured, the kind of mold being made, and the mesh size of the sand, The binder is measured as a percentage the weight of the sand; the catalyst is usually 20% of the weight of the binder. Consult the manufacturer for the correct percentages for their product. See the chart in *Appendix C* for typical alkyd oil binder and catalyst percentages.

As a general rule the following percentages of binder apply for the following conditions:

For aluminum casting piece molds: 1% of the weight of the sand
For cores: 1% binder plus 1–2% wood flour (large cores require more).
For bronze piece molds: 1.5% binder
For iron piece molds: 1.5% binder (for heavy castings add 1% red iron oxide
For melt-out molds (all metals): 1.5% binder, 1% western bentonite clay

Consulting the chart for 100# [45 kgs]of sand to make a melt-out mold, a typical mix would be:

100# [45 kgs] dry silica sand [30 mesh sand]
1.5# [.68 kgs] HA International 18-60 binder
4.8 oz (or .3 #) [.14 kgs] HA International 23-217 catalyst
1.5# [.68 kgs] western bentonite clay.

Carefully weigh the binder in a milk carton, making sure the scale is zeroed out when the carton is on the scale. Drip trays with absorbent material (such as cat litter) are recommended beneath scales and wherever binder and catalyst are drained from containers. Avoid dripping the binder onto the scale; allow the spout to stop dripping before removing the carton. Carefully weigh the catalyst, remembering to account for the weight of the can. Binders are frequently stored in 55-gallon drums and catalysts typically come in 5-gallon pails. While traditional sand mullers are recommended for mixing sand, small batches can be mixed by hand in a wheelbarrow, or a cement mixer can also be used by added several chunky pieces of iron (such as cups from previous castings) to help mix the sand and break up lumps.

To mix the sand, make a trough in the sand and pour in the binder. Cover the binder with dry sand until no more binder comes up through the sand. Allow the binder container to drip empty. Mull the sand for 4–5 minutes. If you are adding dry ingredients to your mix, such as wood flour or clay, add them after the binder is mixed; mull for another 4 minutes. When the muller has stopped, make another trough and add the catalyst, allowing it also to drip out. Cover the catalyst and mull for another 4 minutes. In hot weather the sand will set more quickly; mulling also heats up the sand.

The mulling process not only coats each sand grain with the resin and catalyst, but it conditions the sand as well. **Do not put your hand in the muller while it is on! Be sure tools, weights, or other objects are not in the muller or in the sand before turning the muller on.** Note the time when the sand comes out of the muller; you have approximately 20 minutes of bench life (working time) to complete your molding.

Following your mold making, clean the muller with the wire brush making sure that all excess sand is swept out. Sand and binder will build up on the muller blades (or in the case of a cement mixer on the sides and stationary blades) unless the sand is cleaned.

Making Melt-out Molds

The "melt-out" process is an effective way to cast wax patterns that are relatively simple; it is more difficult to ram up intricate patterns, but with practice and care, even these patterns can be cast in this way. Wax patterns should be sturdily constructed and well sprued and vented. Sprues and vents should extend at least 3" above the pattern. Wax cups are not necessary, as sand cups and vents will be glued on following the melt-out process. If possible, wax the sprue and vents down to a piece of cardboard the same size as the mold. This method helps hold the pattern in the correct position during the ramming of the mold. Cores should be positioned so the core opening or "neck" is up. Cores with small necks should have core pins. Use nails, pushing them through the wax so the heads protrude halfway. Core pins should not line up on a plane, as the mold has a tendency to crack along this line.

Patterns which are 18" or larger must have a run-out attached to the bottom of the pattern (protruding up when the piece is rammed with the sprue and vents down) so that the mold can be melted out right side up and not be rolled over. The run-out should be made from a 3/4" square piece of wax 3" [7.5 cm.] long.

Measure the height, width, and thickness of the pattern with sprue and vents attached. The flasks must be constructed so that there is 2" [5 cms.] of clearance on all sides of the pattern and 3" [.75 cm.] on the bottom. Flasks should be constructed with 1" x 3" [2.5–7.6 cms.] pine. It is recommended that you do not make flasks that are higher than 3" [7.6 cms.] as it becomes difficult to ram down into high flasks. Nail flasks securely together; the height of the stacked flasks should equal the height of the pattern plus 3" [7.6 cms.]. Find a bottom board to fit underneath the flasks. All molds should be rammed up on bottom boards to facilitate easy moving and lifting. Bottom boards should have cleats underneath to facilitate picking up the board. Stack the flasks up and then measure the inside height, width, and depth or the cubic inches inside the flask. One cubic foot or 1728 cubic inches = 100 pound of sand [27000 cubic cms.= 45 kgs]. Consult Appendix D for ease in calculating sand amounts more than or less than 1 cubic foot. Remember the amount of sand needed to fill the flask; disregard the area of the pattern.

Molds weighing more than 100 pounds [45 kgs] should have lifters in the mold. Cut two pieces of thin-walled electrical tubing so that they fit inside the flasks. Position the tubes so that they clear the wax pattern by at least 1-2" [2.5–5 cms.] and are on each side of the mold toward the middle of the mold. Check to see that the tubing fits closely; remove and save for use during ramming. Steel rods can then be inserted to handle the mold after it has set. With thick molds you could drill into the mold after it is set and then insert short steel rods.

The wax pattern should be free of grease, dirt , dust, or silicon. Weigh the wax pattern and note the weight, remembering that 1 pound of wax = 3 pounds of aluminum, 10 pounds of bronze, or 10 pounds of iron [.45 kilos of wax= 1.36 kilos of aluminum, 4.55 kilos of bronze, or 4.55 kilos of iron]. Next prepare the mold facing by mixing by volume 50% zircon and 50% graphite (also called "plumbago") and add isopropyl alcohol to make a sprayable or paintable consistency. Keep the mold wash in a covered container. Brush or spray the mold wash over the entire pattern including sprue and vents to a thickness of approximately 1/32" [.08 cms.] (or until "fingerprint" detail is obscured by the mold wash). Adding a small amount of shellac to the mold wash will help the mold wash adhere to the wax. Brushing a mixture of 50% shellac and 50% isopropyl alcohol to the wax before spraying or painting on the mold wash will also work.

The mold wash acts as a thin refractory shell that holds the detail of the wax pattern; this thin shell is backed by the sand which holds the form of the pattern in the mold. If the mold wash is sprayed, spray only near a ventilation system or outside. Wear a dust mask and be sure a ventilation system is on. The mold wash will dry while you are mixing the sand. Do not allow mold washed pattern to sit around for more than one day as dust will settle on the washed pattern and cause a separation between the mold wash and the back-up sand.

Mix the self-set sand as per procedure. Riddle about 25–50 pounds [9–18 kgs] of sand (to break up lumps) from the wheelbarrow on to a clean spot on the floor or on the molding bench next to your mold. Set wax pattern affixed to the cardboard on the bottom board. Place the first flask evenly over the pattern. Add the riddled sand and tuck the sand firmly and evenly around the sprues and vents. Ram the sand into the corners and sides of the flask and fill to the top of the flask. Add the next 3" [7.5 cms.] flask and continue ramming the sand around the pattern. Continue adding the sand, carefully tucking the sand around the wax pattern. If the piece has a core, ram the wax until the piece is covered halfway and then ram the core up to the neck of the core. When ramming, do not create a smooth plane of sand, but leave the surface rough so that the next added layer of sand will stick to the previous layer. Avoid getting dust, graphite, or parting compound between the layers of sand. Be sure that you ram depressions and details well and that the corners of the mold are firmly rammed. Poor ramming results in a burned-in surface on the casting.

When the mold is half rammed, add the lifting tubes (if needed) and ram the sand over the tubes. Continue ramming the sand and adding flasks until the piece is completely covered. Strike off the top of the last flask evenly as every mold must have a flat top and bottom. If you are making several molds, mark the mold so that you know which pattern is inside as all the molds begin to look alike. After the mold has set (allow at least 12 hours to roll the mold over), remove the flasks and sand the top of the mold flat to reveal the wax sprue and vents. Fit cups and self-set sand vents on the mold and mark their position. Save these pieces until the melt-out process has been completed; they will be glued on just before the casting of the mold. Using an awl or other sharp tool mark your name, weight to be poured, metal, and thickness or thinness of the pattern on the top of the mold

Patterns with flat bottoms and large cores can be rammed as a "roll-over melt-out." After spruing the pattern and spraying it with mold wash, the pattern is placed directly on the bottom board with the flask around it and rammed up. When the sand has cured, roll over the mold, and cut 2–3 undercut channels on the bottom face of the mold (making sure they are far enough from the pattern). The undercuts will lock up the second half of the mold and help keep the two halves from separating during the melt-out and the pour. Add another flask to the top of the mold and then ram the core and remaining section of the mold. When the sand has set up, remove the flask and band the mold, making sure that the bands do not interfere with the cups and vents to be glued on following the melt-out.

When dealing with resin-bonded sands, it is important to remember the 18° F rule (10° C). For each 18° F increase in temperature the resin will react twice as fast. The reverse is true: for each 18° F decrease in temperature, the set time will be slower. This can significantly affect the work time and strip time of no-bake binders and can affect the properties of molds and cores. The slower a resin binder is cured, the greater the core's strength and resistance to humidity.

Melting Out the Molds

The melt-out should be organized so that the kiln is reasonably full. Small individual molds may be melted out in small electric ovens. Large molds with a lot of wax in the patterns should be loaded in a gas kiln so that they are on the burner side of the kiln. Smaller molds or molds with less wax can be loaded on the flue side or stacked on top of the larger molds. Each mold must have a wax pan underneath to catch the wax. Place the wax pan (old cake pans work well) in the kiln and place 3–4 waxy bricks in the pan and position them so that they are not underneath the sprue and vents when the mold is set on them. Molds with run-outs are melted out right side up and smaller molds without run-outs are melted out upside down.

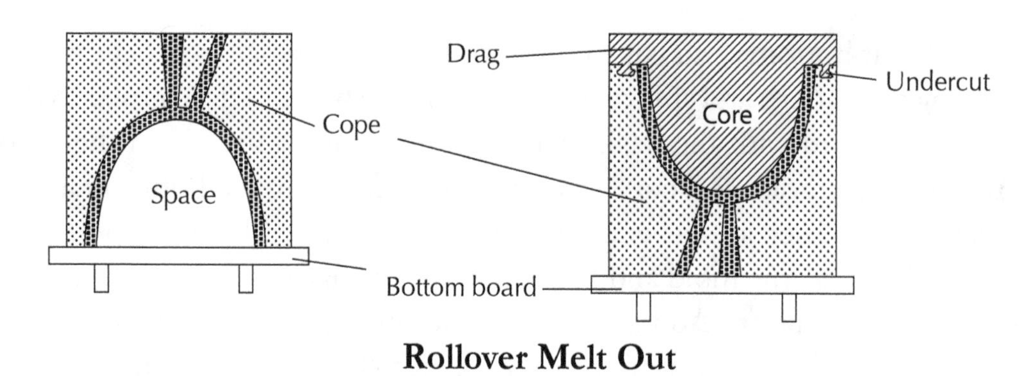

Rollover Melt Out

When all the molds are loaded into the kiln, secure the door. Push the ignition button and wait until the kiln goes through its ignition sequence. When the "Ignition" and "On" lights have lit, set the dial to 300° F [148° C]. Run the kiln for 6 hours at 300° F and then turn the kiln up to 350° F [176° C]. Leave the kiln at this temperature for an additional 18 hours. The kiln should remain on until just before casting. Large molds should be buried in a pit or flasked up with damp dirt, while small molds can be poured directly on the floor. As the molds are removed from the kiln, the wax that has drained into the wax pans should be poured into a pan or bucket to be recycled. Molds that are cracked should be banded and the cracks sealed up with Sairset or other refractory cement. Molds should be organized so that the tops of the molds are at the same level. Remember to organize the molds so that they all are pourable and so that the thin-patterned molds are grouped together, followed by the medium-patterned molds, with the thick-patterned molds at the end of the line. The thin pieces will be poured first while the metal is the hottest and the thickest pieces are poured when the metal has cooled down a bit. (For further information, refer to the "Metal Theory" section in Chapter 8).

After the pour, allow the metal to cool about 1 hour. To begin the shake-out process, break the molds open by gently hammering the molds on their flat faces until the sand fractures in large chunks away from the metal. **Wear safety glasses, gloves, and dust mask during this procedure**. Avoid hitting the casting as it is still soft and the hammer will mar the surface. Shovel the mold chunks and all other shake-out material into the a proper waste container. Most no-bake mold material is classified as hazardous waste and must be separated from the other trash. Sand can be reclaimed by heating it to about 800° F [426° C] for 24 hours to burn off the resin. This may not be economically feasible for many sculptors.

Making Piece Molds

Most wax and foam patterns are destroyed prior to or during the casting process, as the name "lost-wax casting" suggests. In the piece mold process, the mold is made in such a way so that the pattern can be removed without destroying it. If your goal is multiple images, then piece molding is the method to use. Piece molding is generally an easier and shorter process, because there is no need to put the molds into a kiln to eliminate the wax. However, sometimes there is just no other way to make the piece other than with the lost-wax method. The choice of any molding system is to produce the casting in the most efficient and cost effective manner.

Consult chapter 5 for a discussion on how to prepare your particular pattern for casting. Very rigid patterns, such as those made from wood or metal, must be draftable, that is their vertical surfaces must be tapered to allow the pattern to be easily withdrawn from the mold. A taper of 1–2 degrees will work for green sand molds, but patterns that are being made in self-set sand should have a taper of 3–5 degrees. The deeper the "draw" (distance to remove the pattern from the mold), the greater the degree angle should be. Most patterns are set up on pattern boards, or in the case of split patterns, on a follow-board (a removable pattern board that registers the pattern). Loose patterns can be molded by making an initial "false piece," such that the first piece mold made (and, of course, the subsequent pieces) can be removed from the pattern. An undercut in a piece mold section will mean that you cannot remove the sand piece from the pattern without breaking the mold piece. Try to design your pattern without undercuts, or in the case of a found pattern, be sure to false piece it in such a way as to avoid an undercut section.

To false piece a loose pattern, first examine the pattern carefully to determine the number of pieces needed to mold the pattern. If the pattern is already a cast object, it may show the part lines made during the molding process—this is often true of mass-produced objects or objects not finished off particularly well. In the case of a two-sided object, such as a ball, set the object on a bottom board and in a flask so that the flask is even with the part line. Now using Petro Bond sand or green sand fill in around the pattern until the sand extends in a plane around the part line. Smooth the sand with a slick, putty knife, or other broad knife. A draftable portion of the pattern should now be sticking out of the Petro Bond sand. In the case of multiple-sided patterns, embed the entire pattern except for the first draftable section in the Petro Bond sand.

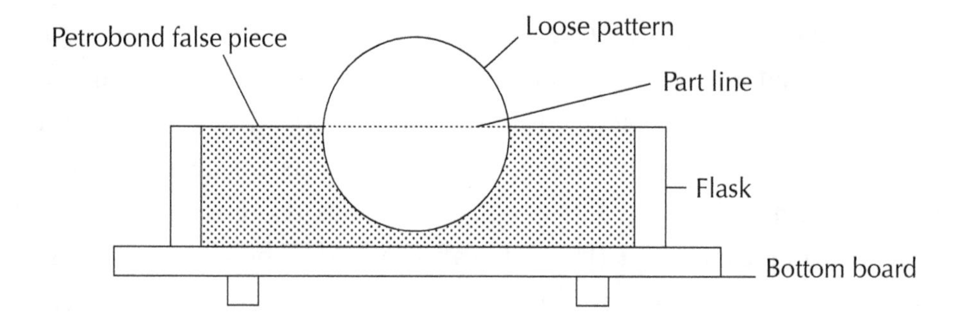

Construct the flasks necessary to mold the pattern, making sure that they are securely nailed. Be sure that you have allowed the minimum distance of 2" between the pattern and the flask and have allowed at least 2" [5 cms.] between the bottom and top of the flask and the pattern. Measure the cubic inches inside the stacked flasks and consult the chart in the *Appendix D* for conversion to sand weight (1728 cubic inches= 100 pounds of sand). Mold sections weighing more than 100 pounds [45 kilos] should have lifter tubes embedded in the sand. Be sure to allow for at least a 1" [2.5 cms.] clearance between the pattern and the tube and at least 2" [5 cms.] between the tube and the flask wall. The larger the mold section, the deeper the tubes should be embedded in the sand. Very large pieces and long thin sections should have welded reinforcement bars (called "gaggers") embedded in the sand.

Cores can present special problems for the molder. Large cores more than 50 pounds [23 kgs] should be made with at least 1% wood flour so that they will be compressible during the shrinkage of the metal around them. Large cores must also be made with some sort of lifting device so that they can be easily handled. To save sand on large cores, the center of the core can be filled with used and already set-up mold chunks, making sure that there is at least 2" [5 cms.] of new sand between the wall of the core and the used material.

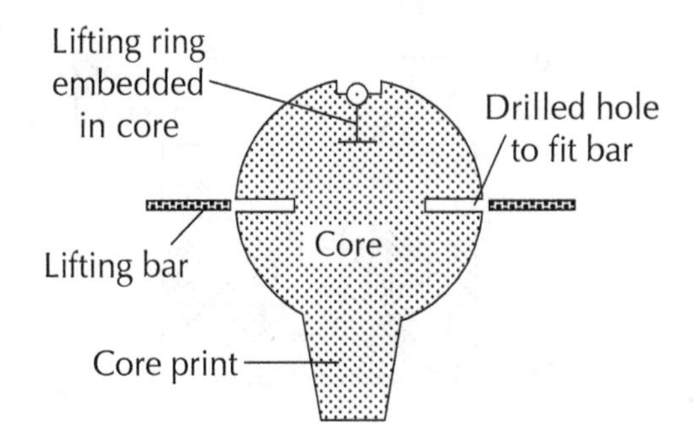

Cores must also register into one or more of the mold sections with a "core print." Cores are sometimes made in two halves and then glued together with core paste prior to their insertion in the mold. Cores can be made directly in the mold, allowed to cure, and then removed. To facilitate the removal of a core rammed directly into the mold, rap the mold with a rubber mallet following the ramming of the core and before it has set-up; the rapping causes the core to compress slightly making it just a bit smaller and easier to remove. Always make score marks before the removal of the core from the mold, so that core can be replaced in exactly the same position. The correct metal thickness can be shaved off of the core by scoring lines of a predetermined depth into the core and all over the surface. To make a 1/4" [.6 cms.] gauge, take a short piece of hacksaw blade and apply tape so that 1/4" [.6 cms.] of teeth are

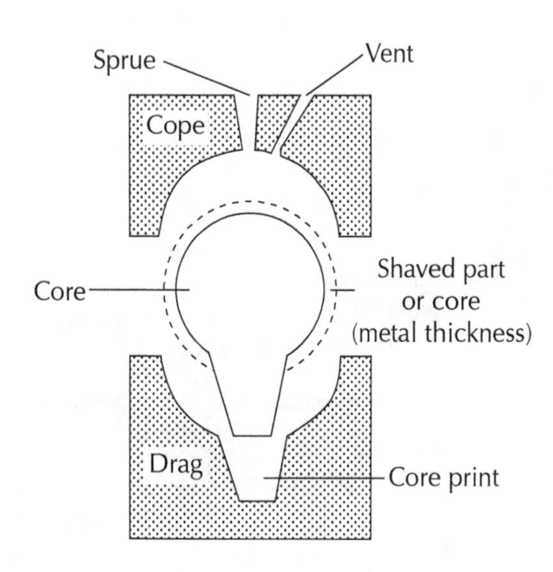

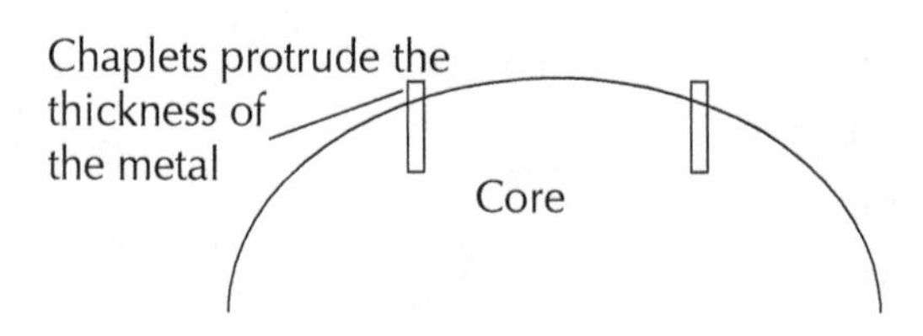

exposed. By sanding the blade into the sand to the tape line, scored lines 1/4" deep will be scored in the surface. Now sand the core with carborundum stones until all the lines have disappeared and you have removed the correct amount for a metal thickness of 1/4".

Large cores should have chaplets. Chaplets are pieces of metal (of the same composition as is being poured into the mold) which protrude from the core at a distance equal to the metal thickness; the chaplets prevent the core from floating up against the cope during the casting process.

When all the flasks, lifters, bottom boards, and patterns are prepared and ready to go, part the pattern by shaking on a thin layer of either talc or graphite, and then mix the appropriate amount of sand. Carefully ram the first half of the mold, paying particular attention to the corners and sides. It is a good idea, especially with patterns that possess a lot of detail, that the initial layer of sand be riddled over the pattern to a depth of approximately 1". The idea is to ram the sand firmly with your fingertips to press the sand evenly against the pattern, then follow this initial stage by shoveling the sand into the flask and ramming further. You will want to ram the mold firmly so that there are no loose areas or gaps in the sand (which the metal will invade giving you a nasty surface or extra form), but not too hard as to take away from the permeability of the mold. Ram the flask or flasks until the pattern is covered with a least 2" [5 cms.] of sand and then strike-off the flask to create a smooth flat top. Mark your mold with your chop or initials, especially if you are working in a cooperative shop, as molds can all begin to look similar. Wait for 12–24 hours and then complete the other half of the mold. A wooden pattern can be more easily removed from the sand mold if the mold is rolled over and the pattern wiggled just a bit to slightly enlarge the sand impression, before the sand has set-up."

Some molds can be rolled over (wiggle or rap the pattern) and the second part of the mold completed right away. Place another bottom board on the top of the freshly rammed mold. If the mold is small, two persons can turn the mold over by holding the two bottom boards together and turning the mold over. If the mold is too large, band the two bottom boards together, roll the mold, cut the bands, and remove the top bottom board. Now cut U-shaped keys in the sand at the edge of the mold.

Cut an uneven number of keys to help you determine how the mold is to be reassembled later—two keys on one side, for example, and one key on an adjacent side. Part the pattern and the mold surface, set on the flasks, and ram the other half of the mold.

One-sided patterns or thick patterns can be made hollow by ramming half of the mold, rolling the mold over, removing the pattern and lining the mold cavity with material the correct thickness of the metal. Liners can be plasticene rolled out to the correct thickness, cardboard, plywood, carpet backing, or foam board.

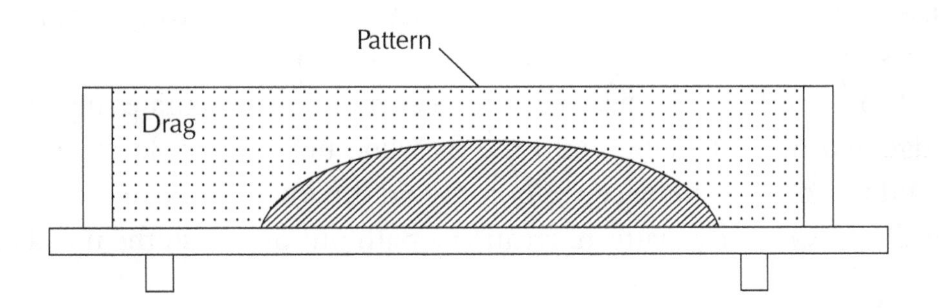

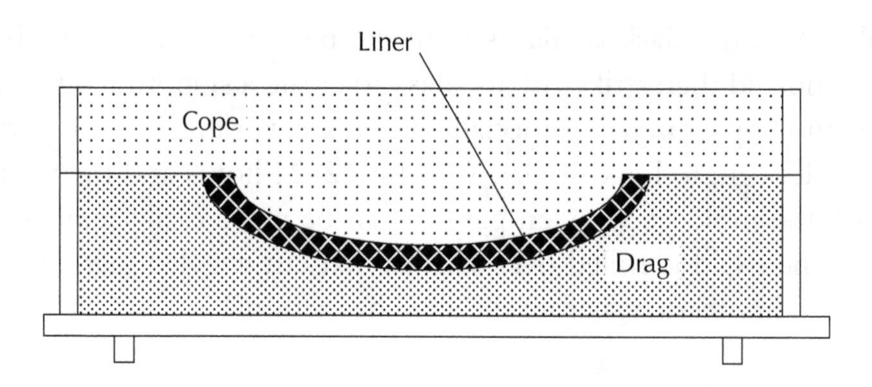

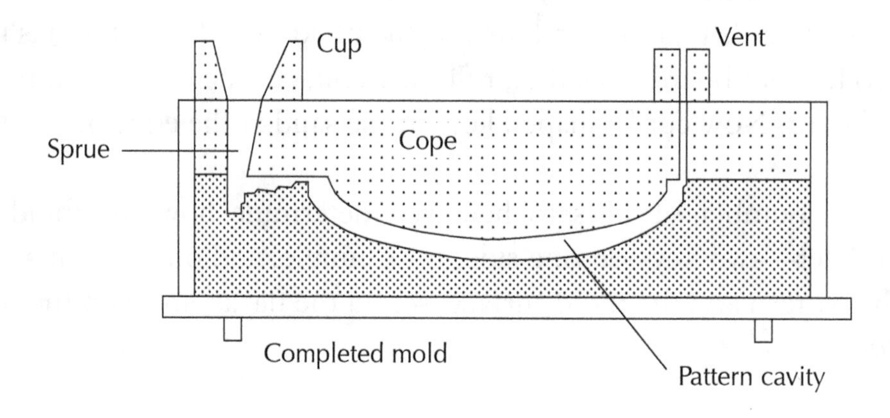

Consult the manufacturer's technical information for the "strip time," which is the time needed for the binder to set the sand sufficiently to be able to remove the pattern. It is probably a good idea to wait at least 12 hours to move the mold pieces or remove the pattern (unless you are experienced, confident, or have an easy pattern to remove). Remove the flasks. Before separating the mold pieces and removing the pattern, use a hacksaw blade to score lines (registration lines) through the part line on all four sides of the mold. Score a different number of lines on each face, so that it will be easy to align the mold pieces during reassembly. To separate the mold pieces, use a wedge and drive it slightly into the part line until the mold cracks along the part line and you can see some separation. Avoid driving the wedge too far as it may penetrate all the way to the pattern, creating a path out of which the metal can leak.

Remove the pattern. Flexible patterns can be easily withdrawn, but rigid patterns may be more difficult. Patterns may have to be rapped with a rubber mallet, or in the case of wood patterns, drill holes and thread in lag screws. Pulling on the lag screws with a pair of pliers should loosen the pattern. Plasticene patterns should not be left in the mold for more than 24 hours as some sand has a tendency to pick up the oil from the plasticene resulting in a weak mold surface. Clean the mold interface by gently blowing it with air from an air gun or brushing with a soft brush.

Additional formal changes can now be effected in the mold through carving, the use of various shaped carborundum stones, sanding, and drawing. Gluing on thin sand pieces into the mold or using the glue itself can achieve negative forms on the castings surface. Holes or other mistaken depressions can be filled through a variety of methods: sand pieces can be carved to fit the defect and glued in place, a mixture of sand and glue can fill the void, or a putty-like graphite material called “slip mud” can be used as a filler. If the spruing and gating system was not part of the pattern, then now is the time to drill the sprues and vents and cut the well, the runners, and the gates. Consult chapter 6 for a general discussion on spruing self-set sand molds. Holes are best drilled with masonry bits followed by mounted carborundum stones to achieve the proper size. Gates, wells, and runners are best cut with rotary carborundum stones. Remember to place all sprues, vents, and gates so that they can be easily cut off and chased following the casting process. For cope mold pieces weighing 25–50 pounds [11–22 kgs] , cut hand holds in the sides of the pieces to facilitate the placement of the mold when reassembling.

When all spruing and gating is complete and you are satisfied with the condition of the mold surface, it is time to apply mold wash to the surface. The mold wash fills in all the spaces between the sand grains resulting in a smoother surface. There are many recipes for mold wash and many commercial products available. A simple and reliable recipe is to mix (by volume) 50% 200 mesh zircon, 50% graphite, and then add isopropyl alcohol to make a sprayable or paintable consistency. The zircon is a tough refractory and resists the penetration of the metal; the graphite is also a refractory, but in addition, it is a carbon and promotes a reduction atmosphere in the mold during the casting. The alcohol allows the vehicle of the wash to evaporate quickly or be burned off. Brush, spray, or flow the mold wash quickly and evenly on the mold interface. It is a good idea to try the consistency you have mixed up on a scrap piece of sand before committing it to your mold. If you have delicate detail in your mold surface, you do not want to obscure it with too much mold wash. On the other hand, mold wash can be applied thickly and texturally to achieve another surface quality. One can even mark back through the mold wash to create additional raised detail or rub the dried surface with powdered graphite for a very smooth surface. Allow the mold wash to evaporate or light it off with a match. With very large mold surfaces or many applications of mold wash, allow some of the alcohol to evaporate before lighting. If the alcohol burns too long it can break down the bond of the sand grains on the surface.

Be sure that all pieces of the mold are clean by blowing them off gently with an air gun. Sand cups and vents should be fitted prior to assembly. Sand the cup against the top of the mold over the sprue to achieve a good fit and do the same for the vents. Poorly-fitted cups and vent pieces will allow metal to leak, causing the system to lose pressure. **Vents must be the same height as the cup to avoid metal running over the top of the mold and on to the floor as the cup fills.** Remember that vents and risers that are at the cup level also serve to keep pressure on the hot metal as it cools, keeping it firmly in place against the walls of the mold. Prior to assembly, mix dry core paste with water to a mix about the consistency of wood glue. Store the glue in a squeeze bottle with a lid. If the glue has been sitting for more than a day, shake to remix as it tends to settle out. Or use commercially available core paste such as "Green Hornet" or construction glue in a tube applied with a caulking gun. Place the drag piece on a bottom board and glue in the core to the core print (if the mold has a core). Then apply a thin line of glue toward the outside edge of the mold. Place the cope on the drag, taking care that no sand falls into the mold, and make sure that all the keys are properly seated and that all the registration lines are aligned. Glue the cups and vents on the cope. Applying a line of glue around the outside bottom edge of the cup and vent pieces will also help assure that no leakage occurs. Cover the cup and vent holes by gluing paper covers on top. Write the required information to identify the maker, metal, thickness of the casting, and weight to be poured on the paper covers. Paper on the cups, vents, or risers keeps dirt or sand from falling into the mold.

Large molds should be banded. Locate bands evenly and support mold edges and corners with steel or litho plate. The mold can be banded directly to the bottom board.

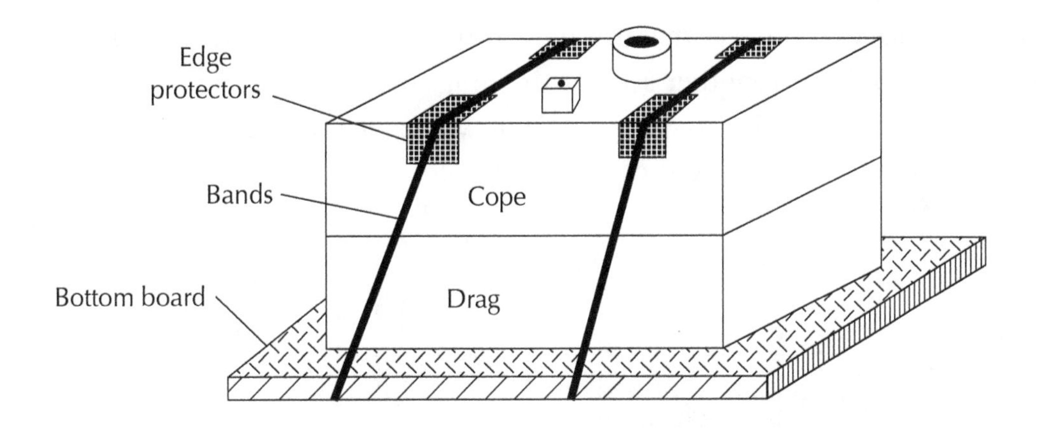

Small molds need not be banded, but can be weighted instead. Banded molds should also be weighted. Multiply the weight of the metal to be poured into the mold by at least 3 to find the minimum amount of weight to add on top of the mold. Molds may also be clamped together with bar clamps or one can use a threaded rod, large washers, and nuts to bolt the pieces together through holes drilled through the mold. (Make sure the holes do not connect to the mold cavity!) If the piece to be poured takes the form of a flat sheet or if the metal has a tall head (a long way to fall) add more weight as more hydraulic pressure is developed. More weight is always better than not enough weight. Sealing the mold outside part lines with refractory cement is also a good preventative measure if you suspect the possibility of a leak through the seam.

Large molds should be rammed up in the pit or in flasks on the pouring floor; use damp sand so that in case of a mold failure and the metal runs out, it will be stopped by the damp sand and not run out on to the floor. Don't forget to wet down the dirt pit before you being placing your molds. Be sure to organize the molding floor so that thin pattern molds (poured hot) and thick pattern molds (poured cold) are grouped together. Molds should also be organized to facilitate easy access to the pouring cups by the pour team. Do not set a short mold in between two taller molds.

Damp sand should also be spread around the molds on the floor to prevent the explosion of the concrete should metal spill on a concrete floor.

When casting several objects that are flat on the back, the molds can be stacked or "ganged" or "booked" together so that the drag of the first mold in the stack is the cope of the second and so on. A common down sprue connects the runners to each mold cavity and a common vent can also be used.

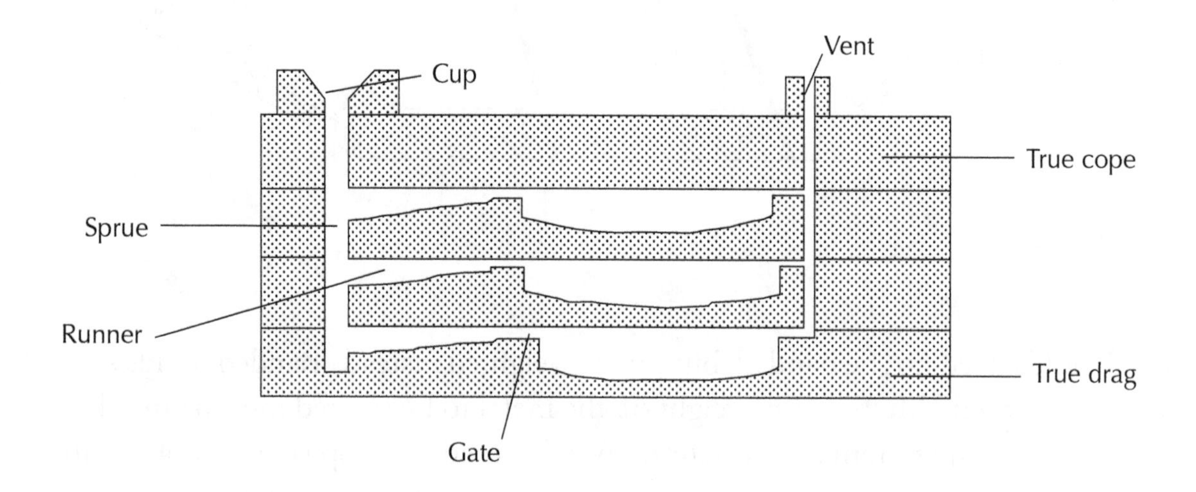

Direct-Built Molds

Most cast metal sculpture is produced by working with positive patterns. It is possible, however, to make a sculpture by building the mold directly without a positive pattern. This method is a good exercise in learning to conceptualize a mold, for one must think carefully about how to fashion the negative space to form the positive object desired. Self-set sand is the easiest way to build molds directly, however it can also be accomplished using green sand, Petro Bond, and clay molds.

To make a simple two-sided sculpture: Make up two identical pieces of sand. The size of the pieces should be gauged by allowing about 2" [5 cms.] of clearance to the top, sides, and bottom. Used pieces of mold material can be reused if the burned sand is removed. Likewise save pieces that are not too badly damaged, especially pieces that have been poured in aluminum, as the low temperature of the aluminum metal does not destroy much of the mold.

Sand the two mold halves together until they fit tightly together (use a figure eight motion for even sanding); no space should be observed between them. Molds can be pre-sanded flat by sanding them against used carborundum cut-off wheels. Now place the pieces together and sand the sides until they are uniform. Score registration lines on all four sides. Open the mold like a book. Using mounted stones, chisels, knives, or any tool that will remove sand, carve the image in the drag half of the mold. Chalk is a good drawing tool to make preliminary images on the mold surface prior to carving, or a cartoon of the image can be transferred to the mold. Remember that everything will be reversed in the casting: "right" in the mold will become "left" in the casting and "in" (depth) in the mold will become "out" and vice versa. Remember a mold is a three-dimensional mirror opposite ("enantiamorph") of the casting you want to make.

The next step is to carve the cope side so that the two images will be aligned. If you have made a cartoon, simply reverse the cartoon so that the former top surface of the drawing is against the cope surface and transfer the drawing. Another method is to carefully fill the edges of the image in the drag with graphite, place the cope on top of the drag, and while securely holding the two mold sections together, invert the pieces. Tap the drag half (now on top) and the image will print on the cope surface. Carefully remove the top half and trace the outline of the image with chalk or an awl. You can now carve the cope part of the mold. The mold can now be completed like other piece molds.

Sculpture can also be made by gluing pieces of self-set sand directly together to form the space that will be occupied by the hot metal. By combining both carving and construction, very complicated pieces can be made. See the work of Julius Schmidt (American sculptor, 1923–2018) for an example of the definitive mastery of this method.

Sodium Silicate Molds

Sodium silicate (Na02 Si02) is a term used to describe chemical compounds composed of sodium oxide (Na02) and silica (Si02) dissolved in water. Sodium silicate is commonly called "water-glass" and is also commercially used as a refractory cement and to harden egg shells. The variety of sodium silicate compounds used in the foundry industry have a lower ratio of silica to sodium making them more viscous and allowing them to react more quickly with the C02 gas or a catalyst used to harden the binder. These foundry binders are identified with the ratio of the silica to the soda; a 2.0 sodium silicate has a proportion of 2 silica to 1 soda.

When the sodium silicate binder is mixed with sand, each grain of sand is coated with a thin film of the binder. When the mixed sand is packed into a flask, small liquid "bridges" of binder connect each and every grain of sand. When C02 gas is passed through the sand or the catalyst begins to work, the binder thickens and forms a rigid bond. Clean sands with a minimum of fines (a high percentage of clay is especially detrimental to sodium silicate molds) should be used. Silica sands with their round grain structure will require less binder than the sharp grain structure of such sands as olivine. The sand should have a temperature of 60–80° F [15–26° C]; cooler sand will retard the set time, while hotter sand will speed up the process.

Additives can be used with sodium silicate to facilitate shake-out, as sodium silicate bonded sand is very hard and difficult to shake-out. Sugar syrup is used with binders that are gassed but will not work with catalytic set binders. Wood flour is a good additive for both categories of binders, especially for cores. Less than 1% western bentonite clay will aid in green strength.

Mixing for Gassed Molds

The advantage of gassing with C02 is that the molds harden almost instantly and can be worked much more quickly. The amount of sodium silicate binder is determined by the fineness of the sand. Between 2.5% and 4.5% binder per 100 pounds sand is added to the sand, dependent upon the grain size of the sand. Mull in additives for 2 minutes (if being used). Add the binder in a similar manner to the alkyd oil binder and mull for 4 more minutes. Mull out into a wheelbarrow or other bin and cover tightly with a damp cloth and a plastic sheet, as the C02 in the air will begin to harden the surface of the sand pile. Patterns should be parted with graphite or talc (N.S.P.); flask the pattern in a similar manner to alkyd oil binders, though bottom boards with holes drilled every 3" [7.5 cms.] will facilitate the gassing of the mold.

Very small molds can be gassed in their entirety, while larger molds should have holes to facilitate the gassing. Using a pointed rod with a diameter of 3/16"–1/4" [.64 cms.], making holes in the mold every 4" [10 cms.] across the surface of the mold and to a depth of which comes within 1" [2.5 cms.] of the pattern.

Roll a C02 cylinder cart to the side of the mold. The tank should be equipped with a flowmeter and valve. Turn on the main valve to allow the gas to flow to the flowmeter. Adjust the flowmeter to 5 CFM (the ball should float at the 5 mark). A rubber hose runs from the flowmeter to an air gun fitted with either a rubber cup or a probe.

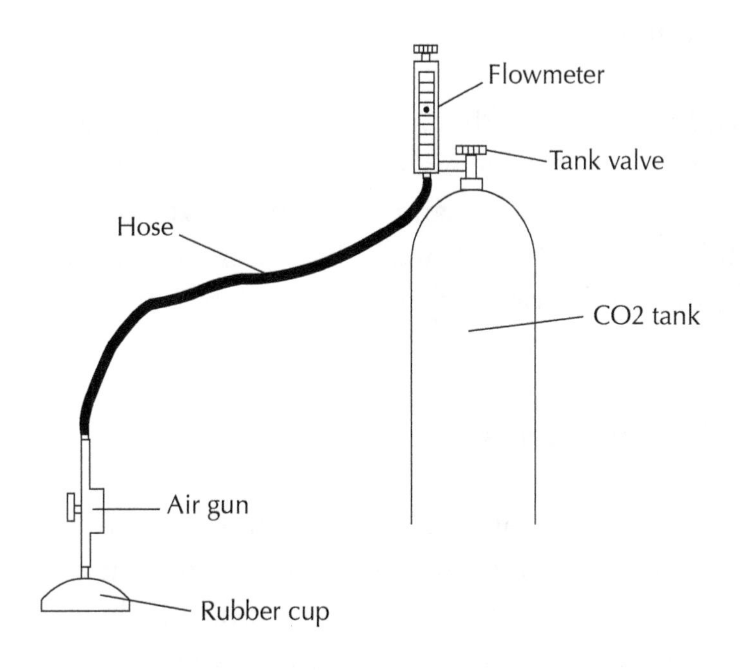

If the mold is less than 5"–6" [12–15 cm.] thick, a rubber cup on the end of the gassing hose is held over an area perforated with the holes, and the gas is allowed to flow for 10–15 seconds. Use a pointed steel rod to make the holes in the sand. The holes should be about 4" [5 cm.] apart and come within 1" [2.5 cm.] of the pattern. Time the gassing and then move the rubber cup to another area of the mold until the entire mold has been gassed. Check to see that the gassed area has become hard; a slight change of color will also be noticed as the water in the silicate is driven from the mold (however, if white crystalline areas appear, you are over-gassing the mold). If the mold is deeper than 6" [15 cm.], use the probe and gas each hole for 10–15 seconds. It is preferable to under-gas the mold, as it will continue to harden through contact with the C02 in the air. Both under-gassing and over-gassing of sodium silicate molds results in weak bonds in the binder and consequently weak sand molds. The molds can be poured as soon after gassing as one can assemble them. Sodium silicate molds should not be allowed to sit around for days at a time as they will absorb moisture and the sand bonds will weaken. Sodium silicate molds can also be baked at 300° F [148° C] to harden. Sodium silicate binders can be used for the melt-out mold technique, but they tend to become very hard. When doing melt-out molding, do not allow the sand to sit too long when adding the layers, as the CO2 in the air begins to set the sand, resulting in cracks in the layers. These defects can leak metal when the molds are poured. Covering the sand with a wet cloth can retard the air CO2.

Sodium silicate cleans up easily with water. Remember to wear latex gloves, as sodium silicate is caustic. Avoid spilling the binder, as when it dries it can fracture into sharp glass-like pieces that can cut your hands and fingers. **In case the binder gets in your eyes, flush immediately with water and then seek medical attention.** Do not wear clothing that has become impregnated with sodium silicate until it has been laundered.

Mixing for Catalytic-Set Molds

Follow the same procedures for handling the sand as outlined in the discussion about the alkyd oil resin. Sil Bond uses sodium silicate as the binder and a liquid catalyst to set the binder. Measure the binder first; normal binder requirements are from 2% to 5% of the weight of the sand. With 30–40 mesh sand 4% binder seems to be hard enough, but consult the manufacturer as to the appropriate amount for the sand used. Next measure the catalyst; the catalyst is always 10% of the weight of the binder, regardless of the speed of the set time.

(Catalysts are available with work and strip times from 1 minute to 70 minutes). Using a catalyst with a set time of about 30 minutes usually gives one sufficient time to make most molds. Consult *Appendix E* for the proper amounts to weigh out.

Consulting Appendix E, a typical mix for 100# of sand would be:

100 #sand
[45.5 kilos]

4# Sil Bond binder
[1.82 kilos]

6.4 oz catalyst
[.18 kilos]

Note: # = pounds

Remember to wear safety glasses and gloves when mixing and handling the sodium silicate binder and catalyst.

In the muller, add the catalyst first and mull for 4 minutes (this is the inverse of resin-based self set binders). Make a trough for the binder, add the binder, cover with sand, and mull for 4 minutes. Mull the mixed sand into a wheelbarrow and use immediately Sodium silicate sands have a tendency to set faster on the surface as the C02 in the air accelerates the set; unused sands may be crusty on the surface. Note that water-based core washes are not recommended for sodium silicate molds; use alcohol based washes only. Use the sodium silicate sand to mold in the same way as for resin-bonded sands. The mold should be allowed to stand for 24 hours before removing the pattern or working with it.

Tools can be cleaned with water. Use a counter brush and a wire brush to remove excess binder and mixed sand from a muller or cement mixer.

Remember that if the mold you intend to make is a simple piece mold, then sodium silicate is the way to go. It makes for a less smoky environment and it produces less hazardous waste.

Standard Investment

Standard investment is a historical and much employed method for casting wax patterns. It is especially useful for large and complicated waxes. Its disadvantage is that it usually contains silica flour in its recipes (silica flour is a hazardous material to the lungs and must be carefully handled) and it must be carefully burned out for a 2–3 day period. There are, however, probably as many standard investment recipes as there are practitioners of the technique. All recipes make use of plaster as the binder, water, and some kind of refractory such as sand, crushed brick, day, grog, vermiculite, luto (crushed and screened used investment), or silica flour.

Mixes containing silica flour or luto (containing silica flour) should be dry mixed outside or they can be mixed inside by carefully placing materials in a 55 gallon drum with a closed lid. Secure the lid and then roll the drum around the floor, turning it over several times, until the materials are well mixed. Dust masks must be worn during any mixing and cleaning activities. **Inhalation of silica dust or flour causes an incurable lung disease called silicosis.**

The author has tested following recipes (all measurements by weight):

Cheap Mix 1
1 part No. 1 molding plaster
1 part 60 mesh silica sand
1 part luto

Cheap Mix 2
1 part No. 1 molding plaster
2 parts luto

Cheap Mix 3
1 part No. 1 molding plaster
2 parts 60–80 mesh silica sand

Strong Mix
1 part No. 1 molding plaster
1 part 30–40 mesh grog
1 part 200 mesh silica flour

Wax patterns should be securely assembled with a wax cup attached to the down sprue and vents equal in height to the cup. Most wax patterns for standard investment are poured with the bottom pour system using adequate runners and gates to feed metal to the mold cavity. It is advisable to wax the cup and vents securely to a piece of cardboard so that the pattern will stand up on its own. Hollow pieces must have core pins and the core opening should be such that the liquid investment material will run in the core and fill it up without trapping air. Core pins can be common nails and should be pushed through the wax so that one-half of the nail sticks out of the wax. Distribute the order of core pins so that none line up along a plane, as the mold will have a tendency to crack along the line. Patterns larger than 18" [45 cms.] should have a run-out vent (3/4" or 1.9 cm. square of wax, 3" [7.6 cm.] long) attached to the bottom of the pattern so that the mold can be burned out in an upright position, avoiding the need to roll the mold over after the burn-out. Molds smaller than 18" [45 cms.] are burned out upside down. Paint a mixture of 50% shellac and 50% alcohol over the wax so that the liquid investment will not resist the wax and contribute to bad surface detail.

Next make a cylindrical cage of expanded metal lath so that the pattern has a clearance of 2" along the sides and 3" at the bottom. Overlap the lath by 4" and wire securely at top, middle, and bottom. Be careful when cutting and handling the lath as the sharp edges can cut your hands. Cut the tar paper to form a jacket that goes around the cylindrical cage one and one-half times. Wire the tar paper at the top, middle, and bottom using soft stove pipe wire with loops in the wire every 5"–6" [12–16 cms.]. Use pliers to twist loops to securely tighten the wire. Mix a small amount of plaster and plaster the cage to a piece of tar paper on the floor; be careful not to bump, move the cage, or dislodge the plaster after this step, as the cylinder will leak when the investment is poured in.

Determine the volume of the cage; for 5 gallons [18 liters] of investment you will need just over 2 gallons [7.5 liters] of water. Add water to a plastic mixing bucket. Add the dry ingredients to the water, stirring it constantly or use a mechanical mixer. Mix thoroughly to achieve a smooth, creamy consistency about like a thin malted milk. Too much water gives you a weak investment that will crack and fall apart during the burn-out process; too much investment material will set up too quickly and not flow around the pattern. If the pattern is set up on a cardboard, place the cardboard in the cage with even spacing around the pattern. Carefully pour investment down the side of the cage until the cup, sprues, and vents are covered. Now splash, flick, or using a soft brush cover the pattern with a thin layer of investment to get the detail covered, paying particular attention to deep valleys and intricate detail. Then quickly pour in the rest of the investment, making sure that the flow does not break the wax. (Some people hold their hands above the pattern and have another person pour the investment into their hands.) Top off the investment right to the top of the cage—do not leave any of the expanded metal lath sticking up beyond the investment material. Agitate the investment material gently or gently tap on the outside of the cage to loosen any air bubbles that are along the sides of the wax pattern. If the pattern has a core, fill the investment up to the bottom of the core, then pour it into the core to fill it up. (Remember that you sprued the piece so that the core opening was up for the mold making procedure.) Then top off the rest of the mold. The mix should set up in about 10 minutes.

Allow the investment to set up and then remove the wire, the tar paper surrounding the cage, and the tar paper on the floor. Roll the mold over, cut out the wax cup and trim the top of the mold so that it is even. If the mold has a run out, trim the wax flush with the bottom of the mold before rolling the mold. Now using a mixture of red iron oxide and water, paint the required identifying information on the top of the mold: name or identifying mark, type of metal, weight of metal to be poured onto the mold, and the thickness or thinness of the pattern. If molds are to be stored for more than a day before the burn-out, wrap them in plastic so that they will not dry out; otherwise; they may go right into the burn-out kiln as soon as they are set up. Adding a bit of dry clay to standard investment dry mix will help the mold retain water if it needs to sit for more than several days.

The Burn-out Procedure

Wax pans are not needed for the burn-out, as they are for the melt-out method, because the wax is burned and not recovered. Most molds are burned out so that they are upside down, except those with run-outs. Set up the molds so that the larger molds are at the bottom and the smaller molds are on top of the larger ones. Molds resting on top of other molds should be raised up with pieces of broken kiln shelves; do not set a smaller mold so that its wax will drain into a larger mold that is being burned out right side up. Arrange the molds so that you are able to see the cups and vent openings of a few molds—this will allow you to visually judge how the burn-out is going. Secure the kiln door. Be sure peep-hole bricks are in a position so that you can pull them out when the kiln is firing. **Be sure all ventilation systems are on before firing the kiln.**

The burn-out takes place in two main phases. In the first phase, the water is being driven from the molds and the wax is being evacuated. It is important for the molds to go into the kiln wet, because not only does the water carry the heat into the mold, facilitating the melting of the wax, but as the wax is expanding as it melts, the water turns to steam and leaves the mold, equalizing the pressure on the mold. For this first phase, set the temperature of the kiln at 450° F [232° C] A schedule should be established so that the kiln is checked every 4–5 hours during the two-day burn-out. (Very full kilns or kilns loaded with large molds may take three days or longer.). During this first phase you will be able to feel the water coming out of the kiln with your hand (or it will smell damp); another way to check for the presence of water is to hold a cold piece of steel or mirror next to the door crack or the upper peep hole—water should form on the cold metal or mirror. You will also be able to smell melting wax.

The second phase of the burn-out begins when all the wax is melted out of the molds and all the water is out of the kiln. The kiln will no longer smell damp or the cold metal check of the kiln will show no water present. The kiln should also smell very waxy at this point. Now begin to raise the temperature of the kiln 100° F [38° C] per hour until the kiln is raised to a temperature of 1100° F [593° C] and held there for 12–36 hours, depending on the size and number of molds. In this second stage the wax that has dripped out is being burned and the wax that has been absorbed by the mold is being changed into carbon and then eliminated with the hot gases in the atmosphere of the kiln. During this phase you will notice the black carbon coming out between the cracks in the door and out the upper peep-hole. Watch this carbon carefully. When the carbon begins to burn away from the crack and the kiln begins to smell hot and clean (rather than hot and smoky), the molds are burned out. You can observe this process by looking in the peep-hole. **Be sure you are wearing gloves, glasses, and a face shield when you open the peep-hole as the blast coming out of the kiln can burn your face!**

You should be able to see flames burning from the cups and vents of the mold in the early stages of phase two, carbon forming in the cups and around the vents in the later stages, and clean dull red molds at the end of the stage with no carbon visible in the mold. Be sure to visually check the lowest molds as they will be the coolest molds.

When you are certain that the molds are burned out, begin turning the kiln down 100° F [38° C] per hour until the kiln registers 500° F [260° C] Then shut off the kiln and close the flue door. Keep the kiln shut until you are ready to cast. Open the kiln and use the mold handling devices and a hoist or hot gloves for small molds to quickly set the molds in the pit. Be sure that a solid footing beneath each mold is layered with a fluffy 1" [2.5 cms.] of sand. Set the molds into the fluffy layer and then gently turn them back and forth to set the mold securely. For the molds that have run-outs, a 3/4" [1.9 cms.] piece of self set sand about 2" [5 cms.] long should be cemented into the run-out with Sairset or other refractory cement. If you have run a proper burn-out, the cups and vents should be white and free of carbon and there should be no checks or cracks in the molds. Cracks running vertically can be wired or gently banded shut and cemented with Sairset or other refractory cement. Molds with horizontal cracks should be also be cemented and weighted when poured. Molds should be securely rammed using the metal rammers so that the pit sand is compacted tightly around the mold. Cover the molds with aluminum foil to prevent sand from falling into the molds. If sand or pieces of investment fall in, vacuum them out, but do not hold the hose too long on the mold as the hot air will burn out the motor or melt the hose! It is an advantage to pour the molds hot, especially if there are thin sections. However, the molds can cool off. If they are not poured within the day they should be returned to the kiln to remove any moisture.

Ceramic Shell

Ceramic shell was developed for the precision casting of aircraft parts and has been adapted by artists for the casting of sculpture wax patterns. Ceramic shell is especially good at capturing the detail of intricate waxes because the pattern is dipped into a liquid slurry. Patterns less than 18" [45 cms.] long are easily cast with ceramic shell; larger patterns present more problems. Ceramic shell is dimensionally stable, resistant to thermal shock, and is so permeable as to allow sprue systems without vents, as gasses vent directly through the shell material.

The ceramic shell mold is made by dipping the wax pattern into a slurry of colloidal silica and silica or zircon flour. Randsom and Randolph Primecoat, Shellspen, and REMET are three often used colloidal silicas mixed with the manufacturer's corresponding silica or zircon flour (200 mesh) to form the slurry. After the initial coats to capture detail, the mold thickness is built up by the application of fused silica. Stronger molds can be made by substituting zircon for the silica flour. Zircon is recommended for heavy iron castings and steel castings. *Methods for Modern Sculptors* by Ron Young has excellent chapters on wax patterns, spruing, and ceramic shell mold making.

Most wax patterns can be top gated with a short 2"–3" [5–7.6 cms.] sprue about 1" [2.5 cms.] thick. Wax patterns must be very securely assembled; bad welding of wax joints or weak assembly will result in pattern distortion or breaking when the heavy layers of shell are applied. Make sure there are no cracks anywhere in your pattern; the shell will penetrate even the smallest crack and cause that part to fall off. Cups should be solid wax and have a an S-shaped hanger of 1/4" [.64 cm.] steel rod embedded in the cup. The hanger is used to support the wax pattern during dipping and when drying. Securely weld the sprue to the bottom of the cup. Allow the wax to cool completely before proceeding to the next steps.

An alternative method for spruing is to use the hollow red wax sprues. Weld the sprue to the pattern in the same manner as the solid sprue. Make a hanging device by welding a ring on the top of a 3/8" lag screw, using sticky wax to stick a Styrofoam cup to the red sprue. After everything is cool, screw the lag screw hanger through the middle of the Styrofoam cup deep into the hollow red wax sprue.

This method eliminates the need to melt the wax out of the cup, as the cup stays hollow when the slurry is applied around it. To keep the slurry out of the inside of the cup, cut a paper pattern such that it fits the top of the cup and use hot wax to seal it to the cup.

Weigh the wax pattern (with cup attached) and note wax weight for conversion to metal weight. Brush the entire pattern (including the cup) with a thin coat of shellac and alcohol (50% shellac and 50% isopropyl alcohol by volume). The shellac mixture will allow the slurry to stick to the wax surface. Allow the shellac mixture to dry thoroughly before proceeding. Alternately, you could spray shellac on the wax using a spray can of shellac available at most hardware stores.

If your pattern has a core with a narrow opening (or neck), you will need to fill the core with special core material. Use stainless steel core pins (regular steel rod or nails will not survive the vitrification process) through the wax and into the hollow wax where the core will be. Core mixes are available from each manufacturer.

The core must be thoroughly dry before beginning to apply any of the slurry and stucco coats. For cores with large openings (more than 1" [2.5 cm.]), the slurry and stucco can be applied on the inside. The use of core drying hoses from an air supply or small fan is recommended to be sure the core is dry prior to applying an additional coat. Insert the hose into the core opening or position the shell so the fan blows into the core opening to facilitate drying.

The slurry is kept in constant agitation in a rotating stainless steel slurry tank with L-shaped paddle mixer or with a stainless steel propeller mixer. If the slurry is not continually mixed, the heavier silica flour (or zircon flour) will drop out of suspension and separate from the colloidal silica creating a hard mass on the bottom of the mixer.

Keep the mixer running! If slurry is too thick, do not add tap water. Use distilled water or more colloidal silica. Wear disposable gloves when coating your wax pattern with the slurry. Dried slurry is extremely difficult to remove from tools, floors, and surfaces—it is recommended that you not slop the slurry around.

The first "face" coat is very important because it captures all of the detail on the surface of your wax pattern. Carefully dip, brush, or pour (reuse a paper coffee or soft drink cup) the slurry over the wax pattern while holding it with the cup down or hanging from a hanger over the slurry tank. In most commercial foundries the slurry tanks are large enough for the patterns to be dipped facilitating even coating of the slurry. Large volumes of shell slurry are expensive, so this option may not be practical for small operations. Avoid holding your pattern horizontally. as patterns seem most vulnerable to breaking in this position. Make sure pattern is evenly coated and allow excess to drip back into the tank. Do not get slurry over the top edge of the cup. Waxing a paper cover over the top of the cup will help you avoid getting slurry on the inside of the cup. Avoid spilling or dipping the slurry on the floor or edge of the mixer, where it will dry and create a hazardous dust (see "Dust" section of chapter 4). When excess slurry is dipped off, hang the shell on a rack. Turn on a fan to accelerate the drying of the slurry. Clean brushes in jar of distilled water; allow excess slurry to drip out of any paper cups used and dispose.

Allow the face coat to just get dry (10–20 minutes depending on the temperature and humidity). The slurry will turn a darker yellow-orange when it is dry (if you are using the colloidal silica product that changes color when dry) and will lose its sheen. Apply the second coat in the same manner as the first. Once again avoid covering the top of the cup. When all the excess slurry has dripped off, carry the pattern to the fine stucco box containing the fine fused silica. Gently coat the pattern with the fine stucco so that an even layer adheres to the slurry. Wear your dust mask during all stuccoing operations. When no more fine stucco adheres to the slurry, hang up the pattern to dry. Again the drying process may be accelerated by use of the fan blowing at a low speed on the shells. Drying times also depend on the temperature and the humidity in the air. Open cores and deep depressions may take more time to dry; check these areas thoroughly. Use the core drying hoses to get the cores dry. Surgical tubing attached to a regulated air source works well, as the tubing can be inserted into the core directly.

Apply the third coat in a similar manner and allow to dry. You now have one face coat and two fine stucco coats on your pattern. Should more than several hours elapse between coats, re-wet the previous coat by brushing on a thin coat of the colloidal silica without the added flour. This will assure that the next layer of slurry bonds to the previous coat.

Apply the fourth coat of slurry, allow excess to drip off as before, then stucco with coarse fused silica. Flotation tanks that float the coarse stucco with air work exceedingly well.

The stucco should rise up in the flotation tank when the air valve is on. Wearing your dust mask, simply dip the wet pattern into the tank and the stucco will be evenly coated all over. Plunge again to be sure that no more stucco will adhere to the pattern. The top edge of the cup should now be just beginning to build up, but remember not to cover the whole top surface of the cup or spill the shell into the Styrofam cup (if you are using this method), or you will have to chip and cut the hard dry shell away in order to maintain the opening in which to pour the metal.

Repeat the procedure for coats 5, 6, 7, and 8. Be sure that the top edge of the cup and the cup itself is as thick as the rest of the shell. Build the shell up until it is just over 1/4" [.64 cm.] thick. For most small patterns, this will be sufficient. If your pattern is larger or has flatplanes, you may have to build up several more coats. When the final stucco coat has dried, brush a final coat of slurry over the stucco. This will keep the stucco grains from falling off and makes the shell easier to handle. Allow the shell to dry thoroughly.

If your pattern is a thin relief or has very thick sections with lots of wax, it may be prudent to relieve the wax pressure in those areas, as wax expands when it melts and the expanding wax in thin flat sections or in very heavy sections may crack the shell. Drill through the shell at the thick part with a 1/4" masonry bit until you just hit the wax, or use a hacksaw blade to saw through the shell at the edges of flat thin sections until the wax is just nicked. Try to drill or nick in places with little or no detail. These holes will allow the expanding wax to flow out of the shell before the shell cracks. Plug the holes following de-waxing by applying refractory cement over the holes to seal them up.

You are now ready to de-wax the shell. If you have a solid wax cup, it is best to remove the cup before placing the shell in the flash-out kiln. Place the shell in a wax pan with the cup sloped downward. Using a propane torch, heat the outside of the shell in the cup area until a bit of the wax flows out. Heat all sides of the shell cup, but avoid heating the rest of the shell. Twist the hanger to loosen the wax cup from the shell and pull it out. If you are using the Styrofoam cup method, pull and break the cup away from the shell exposing the wax sprue at the bottom of the cup.

Carry the shell to a flash-out kiln. Most flash-out kilns have a bottom grate that will allow the wax to drain out of the kiln into a water pan located underneath the kiln. Place the shells cup side down in the kiln; bigger shells should be located closest to the burners. It is okay to rest the shells against the sidewalls and each other, just be sure that they will not fall during the de-waxing. Close the kiln door and secure.

Adjust the burners so that the flames are striking the shells. The wax will drip out of the shells and collect in the water pan. When no more wax is dripping, turn off the gas valves to stop the burners. Allow the shells to cool and then open the kiln door. Remove the collected wax from the wax pan after it has cooled.

The shell will have carbon on the inside and perhaps some on the outside. Allow the shell to cool before moving it. Although the shell can be handled at this point, it will not achieve its maximum strength until it is vitrified. Be careful with the shell at this point—cushion it with paper or cloth until you are ready to vitrify. Check the shell carefully for cracks. Cracks can be repaired at this point by applying slurry and stucco over and around the cracked area or with refractory cement. Bad cracks or broken pieces can be repaired with fiberglass cloth and slurry or refractory cement. Allow repairs to dry thoroughly before proceeding to the vitrification process.

While shells can be poured unsupported, it is recommended that they be supported with dry sand in a metal bucket. Select a metal bucket large enough and tall enough to support your shell; put 2" [5 cm.] of dry sand into the bucket and gently set the shell inside. The shell should not touch the sides and the cup should be even with the top or just slightly above the rim of the bucket. Remove the shell. You are now ready to vitrify the shell.

Face shields, gloves, glasses, and protective jackets should be worn when removing the shells following the vitrification process.

Place the dewaxed shells gently into the kiln. It is okay to rest the shells against the side walls and each other, just be sure that they will not fall during the vitrification process. Close and lock the door. Note the time. Most shells should be heated until they are a red-orange color, usually about 15–20 minutes, but the time will vary depending upon the kiln. Check the shells through a peep-hole, making sure you are wearing safety glasses and a face shield. Do not overheat the shells.

When the shells are vitrified, turn off the burners and wait until the color is gone from the kiln. Wearing your protective equipment, remove the hot shells from the kiln using a raku tong or handling them with Kevlar gloves. It is easiest to do this as a team with one person grabbing the hot shell and another person holding a shovel ready to accept the hot shell. Move the shell on the shovel to your previously fire-proof surface. Seal any cracks with refractory cement directly on the hot shell. Cover the cup with aluminum foil and set it in the bucket. The top of the cup should just be above the top of the bucket. Add dry sand to cover the shell. The inside of the shell should be white with no carbon present. Chalk the weight to be poured into the shell on the side of the bucket. The shells should be poured as soon as possible after removal from the kiln. Hot shells will facilitate the casting of pieces with very thin sections. Some founders pour the shells without a bucket without surrounding sand by just leaning them against a rack. If you are positive the shell will not crack or break, you can do likewise.

Following the pour, allow the metal to cool for at least one hour. **The sand around the shells will be hot! Beware of burns from the hot sand. Wear dust mask, safety glasses, and gloves when removing the shell from the casting.** The shell should break off in large pieces when gently tapped with a hammer; hitting the top of the cup is a good way to cause the shell to drop off. Avoid hitting the piece itself. The remaining shell can be removed by wire brushing or sandblasting, but be sure to screen sand when finished to remove shell pieces.

Dispose of the shell in a proper place as the shell is classified as hazardous waste (silica flour). It is recommended that you do not get the shell in a wax working area or in the pit sand.

If you have successfully completed all of the procedures correctly, you should have a defect-free casting with a perfectly reproduced surface texture!

Alternative Shell Mixes

Both Remet and Ransom and Randolph now make products that need not be continuously mixed.

"Shell-spen" is a proprietary product that eliminates the need for constant mixing of the slurry, making it possible to do shell molding in small batches, available from Shellspen International, PO Box 1741, Venice, FL 34284 (941) 951-2545 [www.shellspen.com]. The product is added to any regular colloidal silica and, once mixed, stays in suspension. Most likely, the suspending agent is bentonite clay.

The following formula can be mixed using readily available materials and has been used at several university foundries around the country:

2 quarts [2 liters]
colloidal silica

1 quart [1 liter]
distilled water

6 quarts [6 liters]
mesh silica flour

1/4 quart [.25 liters]
225 mesh western bentonite clay

Dry mix silica flour and bentonite and then add to liquids; alternatively one could mix the western bentonite into the distilled water first, then add the colloidal silica. Stir with power drill and mixing shaft. Test viscosity with #5 Zahn cup; a ten-second drop should give the right viscosity for coating most waxes. This mix should be mixed with a propeller-type mixer about once per day, but does not need to be constantly mixed. Small batches are best. Apply and stucco as per any ceramic shell. This formula tends to make a weaker shell mold, so additional coats may be needed to reach the strength of other shell formulas.

Clay Mold Making

Clay was one of the earliest molding materials and figures in the metal casting history of Africa, Meso-America, Greece, Japan, India and China. Small castings can easily be done using this readily available and inexpensive material. Large castings will probably be beyond the reach of everyone but the dedicated researcher or technician. In Africa, lost-wax casting in silver, gold, and bronze was developed using clay molds. In Meso-America, mainly gold and silver was cast. In China, astonishingly intricate castings were being made with clay piece molds. The Chinese were casting very thin bronzes and iron centuries before the Europeans.

Lost-Wax Casting Using Clay Molds (Meso-American)

Small waxes less than 5" [13 cms.] can be cast using clay molds containing charcoal. The wax pattern should be sprued using a top gate system with connecting threads of wax to any extremities. The main sprue is connected to a small cup of wax with a hooked wire embedded in it (the wire facilitates the hanging of the wax during the drying periods). The wax pattern should be painted with a 50/50 mix of alcohol and shellac which promotes the adherence of the water-based slurries of clay to the wax. The wax pattern is then coated with slurries of earthenware clay and charcoal.

Prepare the slurries according to the following recipes (the amounts are measured by volume):

Slurry # 1:
75% crushed charcoal
25% red earthenware clay
Water to make a creamy slip

Slurry # 2:
50% charcoal
50% clay
Water to make creamy slip

Slurry # 3:
25% charcoal
75% clay
Water to make creamy slip

Coat # 4:
75% clay
25% crushed brick or grog
Water to make modeling consistency

To begin making a lost-wax clay mold in the Meso-American manner, dip the wax into the first slurry so that the clay has a thickness of about 1/8" [.3 cm.], or paint on the first coat with a soft brush. Allow the clay to dry until it loses its watery sheen (soft, but not at the leatherhard stage). The second and third slurries are applied in similar fashion, building up a layer of clay that is approximately 3/8" [.9 cm.] thick. The fourth coat is applied when the clay is also at a soft stage. Model on a thickness of approximately 1" [2.5 cm.] from all points with a slightly thicker application around the cup area. Allow the molds to dry thoroughly. When completely dry, the molds should not feel cold when held to your cheek.

When completely dry, the wax can be eliminated from the molds by placing them next to a wood or charcoal fire with the cups down. Some of the wax can be recovered by scratching a trough in the dirt to collect the wax. With the wax eliminated, the molds are placed in small wood-fired or charcoal-fired kiln, with wax cups down (so that ash and charcoal bits do not get in the mold). The kiln should be fired until the interior of the kiln and the molds reach a dull red color (approximately 1000° F or 538° C). An alternative method of firing is to simply group the molds securely on a fireproof surface and then build a fire over and around the molds. Continue to keep the fire going until the molds are a dull red color The clay should have the consistency of very soft brick or a low-fired flower pot. Do not over-fire the molds as they become too hard and it is difficult to remove the hard clay from the casting. Allow the molds to cool a bit and then pour the molds hot. Molds with very thin patterns should be poured when they are still red hot inside.

Please note: All the formulas in the clay molding section are for making the initial batches of the material. The beauty of clay molds is that the material can be re-used. Following casting, simply crush up the material, re-screen, and use again. Add clay slip to re-wet the material. After many uses, you may need to add additional clay to the mix.

Clay Piece Molds

Simple patterns can be cast using clay piece molds. Clay molds can also be built without the use of patterns by carving the negative area directly. Patterns should be water resistant; graphite seems to work best as a parting compound since it is slightly "greasy." The main problem you will encounter in using clay for a piece mold is the shrinkage of the clay during the drying process. To begin the clay piece mold process, prepare two batches of clay: in the inner coat, the charcoal acts to make the mold very porous next to the metal interface and the wollastonite helps to retard shrinkage of the clay; in the outer coat, grog helps strengthen the mold and again the wollastonite retards shrinkage.

Inner coat:
- 30% red earthenware clay
- 15% 60 mesh silica sand
- 45% crushed charcoal
- 10% wollastonite

Outer coat:
- 30% red earthenware clay
- 30% 60 mesh silica sand
- 30% grog
- 10% wollastonite

The clay should be prepared with the minimum amount of water to make an easily moldable consistency and well kneaded to eliminate any air.

For a simple two-sided solid object (assume split pattern on follow board or a pattern with one detailed side and a flat back), part the pattern with graphite or oil, then make a slab of the inner coat day about 3/8" [.95 cms.] thick. The clay can be patted out on a flat surface or rolled out with a rolling pin. Place the slab on the pattern and press to form the slab over the pattern and trim so that mold extends at least 2" [5 cms.] from the edge of the pattern on all sides. Score the top and side surfaces of the slab and model on another 3/8"–1/2" [.95–1.27 cms.] layer using the outer coat clay.

Turn the slab and pattern over. If the pattern is a split pattern, rap the follow board to loosen, pull it off, and then set the other half of the split pattern in place. Wiggle the pattern slightly to loosen it up from the clay. If the pattern has only one detailed side, rap the pattern board and pull it off. With a split pattern, part the clay and the pattern with graphite or oil first and then also talc. With a one-sided pattern fill the mold cavity with dry sand up to the level of the mold thickness or line with the correct thickness of cardboard, carpet backing, or foam core, then part with both graphite and talc.

Cut shallow U-shaped keys in the first mold half. Now repeat the process of placing the inner coat clay slab on the pattern and parting surface of the first mold half. Model on the second coat day in a similar manner. Score registration lines on the sides through the two mold halves. Place the completed mold on a bed of grog and allow to dry for 2–8 hours (depending upon the humidity).

Open the mold and remove the split pattern or the sand. Using a sharp knife cut a sprue, vents, and a cup; simple direct gating seems to work the best. Further detail can now be stamped or drawn into the mold cavity, or one can wait until the mold is leatherhard to carve even finer detail. Slip made from the inner coat recipe can also be brushed on to give other surface detail. Place the two mold halves together and allow them to dry. When leatherhard, use slip to glue the mold halves together. Allow to dry further until the molds are completely dry and not cold when held to the cheek.

The molds can be fired in an electric kiln until they are soft bisque (cone# 06) or they can be fired in a charcoal kiln until they are dull red in color. Use wire to secure the molds halves and preferably pour them while still hot, about 400–500° F [204–260° C]. Do not over-fire, as the clay becomes very difficult to remove from the casting.

African Clay Molds

Lost-wax casting in Western Africa dates from about the 9th century. The material in this section comes from comparing various books and articles along with extensive fieldwork done in Ghana by University of Minnesota student Bart Cuderman in 1995.

Two methodologies emerge in African casting in regard to the method of metal delivery to the burned out mold cavity: one, an open cup method, similar to most casting, where metal is poured into a cup connected to a top-pour sprue-system, and two, a closed cup method, where brass or bronze pieces are sealed into a cup chamber also connected to a top pour sprue system. In this second method, the wax is burned out with the sealed cup chamber down. The furnace is then brought to a higher temperature to melt the metal in the cup chamber, and the mold is inverted to allow the liquid metal to run into the sprue system and fill the mold cavity. This latter method is similar to techniques also employed in India.

Two kinds of clay molding materials are found in Western Africa: a clay and charcoal method, similar to Meso-American molding, and a clay and dung method. Both clay formulas and casting methods produce good results with small wax patterns.

Ashanti Clay and Charcoal Molds

The wax pattern is prepared with a top pour sprue system. The sprue should be longer than necessary, so that it can be easily handled. The longer sprue can also be stuck in a bed of sand to facilitate drying of the clay layers. The pattern is given a thin wash of a 50/50 shellac and alcohol mixture so that it will accept the first mold coat. The charcoal is prepared by crushing it with heavy hammers and then screening it through a fine (75–100 mesh) screen or riddle. The clay can be locally gathered and either dry sifted or water sifted to remove most organic materials. "Red Art" or other commercial earthenware clay will also work.

Four mixtures are used (by volume) to complete the mold:

1. Primary Slurry
2/3 charcoal
1/3 clay
Water to make thin creamy slurry

2. Secondary Mixture
1/2 charcoal
1/2 clay
Water to make wet, but moldable mixture

3. Tertiary Mixture
1/3 charcoal
1/3 clay
1/3 crushed old molds, fine grog, or fine sand
Water to make moldable mixture

4. Final Mixture
1/3 clay
1/3 crushed old molds, fine grog, or fine sand
1/3 chopped fiber (straw, dried hay, palm fiber, etc.)
Water to make moldable mixture

Dip or brush the primary slurry onto the wax pattern and allow to dry until the slurry looses its sheen. The slurry should be thick enough so that fine detail (such as fingerprints) in the wax pattern cannot be seen. When slurry has dried so that it can be handled, but is not completely dried out, apply the secondary coat by brushing or modeling with the fingers to a thickness of 1 cm. [less than 1/2"]. Allow to dry to leatherhard and then wash with a very thin slip of clay and a model on third coat mixture 1–2 cm. thick [1/2"–1", depending on size of wax pattern]. Begin to build up cup around protruding sprue with this coat. Allow to dry to leatherhard, then apply slip coat, and model final mixture to an additional 1–2 cm. [1/2"–1"] thickness, while completing the cup.

Allow mold to dry thoroughly. Wax may be eliminated in an oven or in a charcoal fire that is not hot (no red coals) to recover wax prior to curing. Cure molds in a simple kiln fired with charcoal until they are dull red in color. Molds should be poured hot, as soon as they are removed from the kiln. Allow the molds to cool after pouring. Save the old mold material to be recycled and used for the next generation of molds.

Bifor Molding

Prepare the wax pattern in the same manner with a top pour system and a longer sprue to facilitate handling and positioning for drying. In Northwestern Ghana the Bifor tribal group uses a clay-like topsoil found in deserted and eroded termite mounds along with donkey dung to make molds for casting. Having no access to termite mounds, we have substituted locally gathered clays and commercial bond clays such as "Hawthorne Bond" or "Red Art" for the termite mound material. Local clays should be either dry sifted or water sifted to remove large organics, rocks, or other non-clay materials. Dried horse manure works the best, although dried cow manure will also work. Crush the manure and then screen it through a 30–50 mesh screen. Use a dust mask when screening dry manure.

Mix 2/3 dried and crushed horse dung with 1/3 clay. Add water to make a wet, moldable consistency that will not stick to the fingers. In Ghana, this mixture is applied directly to the wax pattern by evenly pressing the clay mixture around the wax until a thickness of 1 cm. [about 1/2"] has been built up, leaving the end of the sprue exposed. This takes some skill to make sure the mold material is applied evenly into the detail. For very detailed wax patterns, we have modified this technique by making a thick slurry of the clay and dung mixture and applying it with a brush to a thickness of .5 cm. [about 1/4"] . When this slurry coat has dried to leatherhard, an additional .5 cm. of moldable material is added.

Apply a second coat, leaving the sprue exposed. Allow to dry to a leatherhard state. Build up a cup about 5–7 cm. [2–3"] in height, depending upon the size of the wax pattern and all around the exposed sprue and then continue applying a third coat to a thickness of 1 cm. [about 1/2"] around the entire mold. Allow to dry till leatherhard.

Prepare a fourth coat material by adding chopped and pulverized dry straw to the clay substituting for the dung. The straw will make the final coat very porous. Add enough water to make a moldable consistency similar to the other three coats of material. Now cut up sufficient metal to fill the pattern and the sprue. Remember to weigh the wax before molding: 1 pound of wax [.45 kilo] will weigh 10 pounds in bronze [4.5 kilos]. Metal chucks should be small enough so that they fit well within the volume of the cup you have modeled on the top of the mold.

Build the fourth coat up to the lip of the cup around the mold and to a thickness of 2 cm. [about 1"]. Add the metal to the cup. Prepare a flat patty [2 cm. thick] of the fourth coat material and cover the cup, sealing the edge of the patty to the edge of the cup. Allow the mold to dry thoroughly.

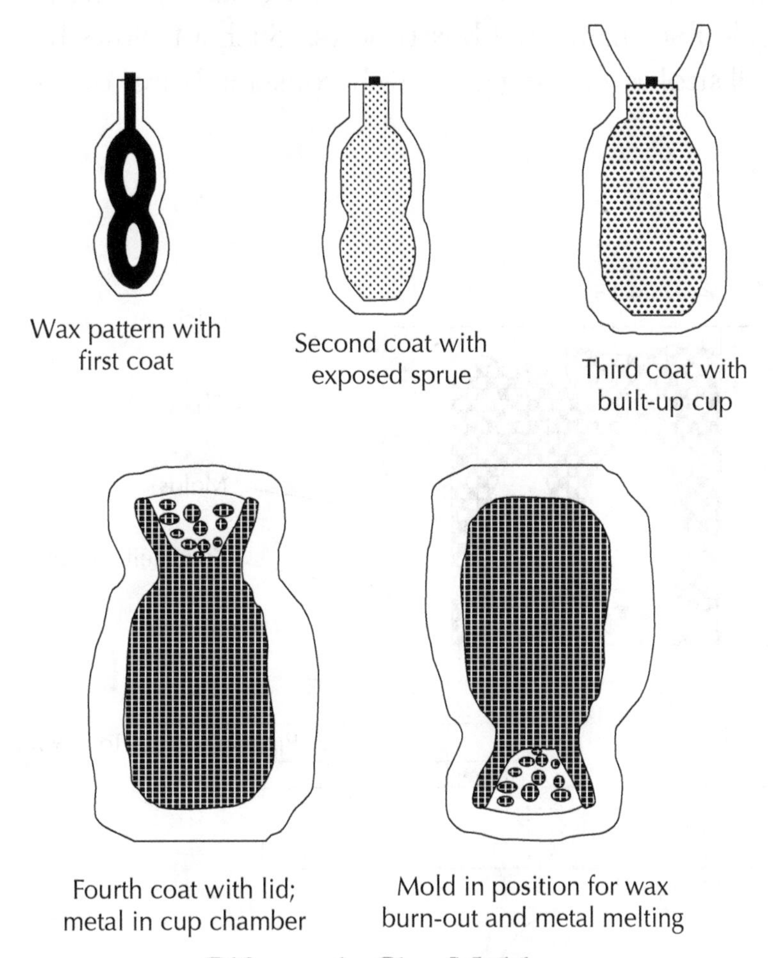

Bifor-style Clay Mold

The molds are now placed in a furnace with the cup end of the mold down. In Africa, furnaces powered by hand bellows and fueled with charcoal are used to bisque the mold and to melt the metal. With a hand-powered bellows, this process may take 4–5 hours. One sign that the molds are ready to be inverted (if red brass or yellow brass are being melted) is that a discoloration of the charcoal to a white or yellow-green hue or signs of white vapor from the fuming of the zinc in the alloy can be observed. A tong is used to pick up the individual molds and turn them over so that the molten metal flows into the sprue and into the pattern cavity completing the casting.

One can build contemporary versions of the traditional "Saa Ngia" ("the Red Furnace" in Bifor). Using bricks, a clay mortar made from sand and fire clay, a metal "pela" or furnace grate, and powered by a small blower and fueled by charcoal, the furnace is capable of curing molds and melting the metal in about 1–1.5 hours. One can also be successful in achieving melting temperatures using a hand-powered bellows modeled after a Chinese double-chambered box bellows. Such a bellows blows on both the push and the pull strokes, creating a relatively constant source of air.

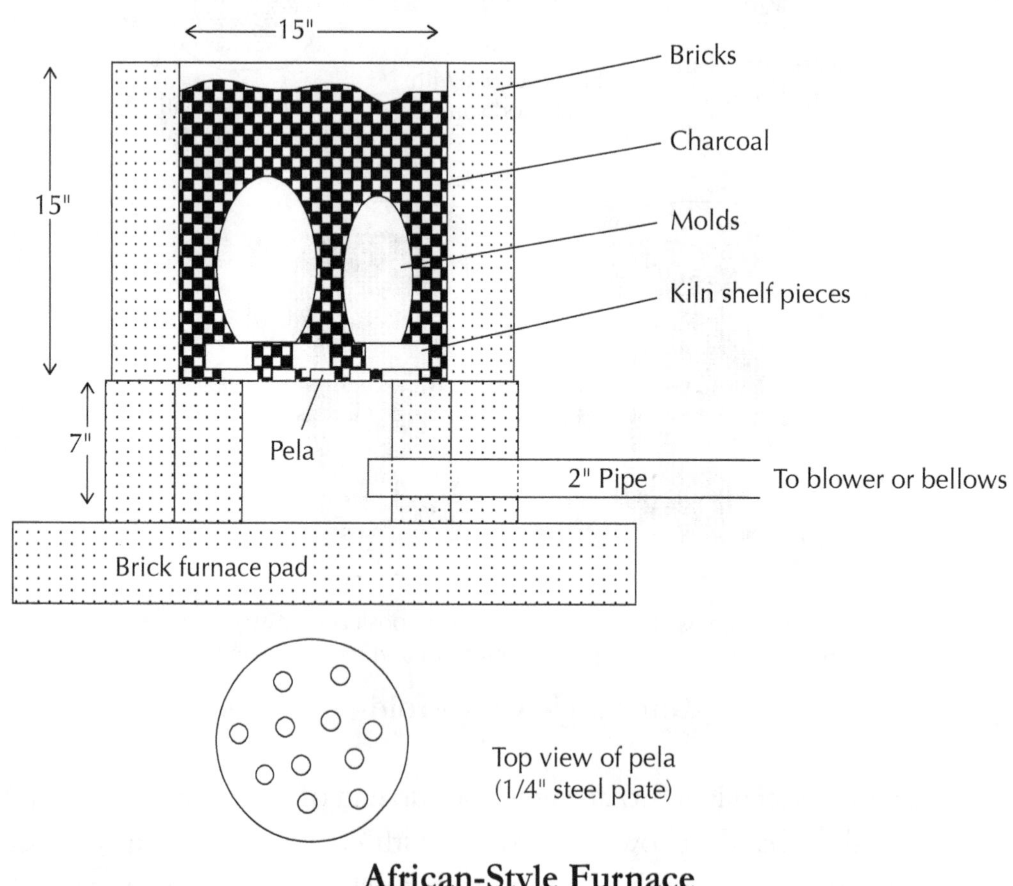

African-Style Furnace

Japanese Clay Molds

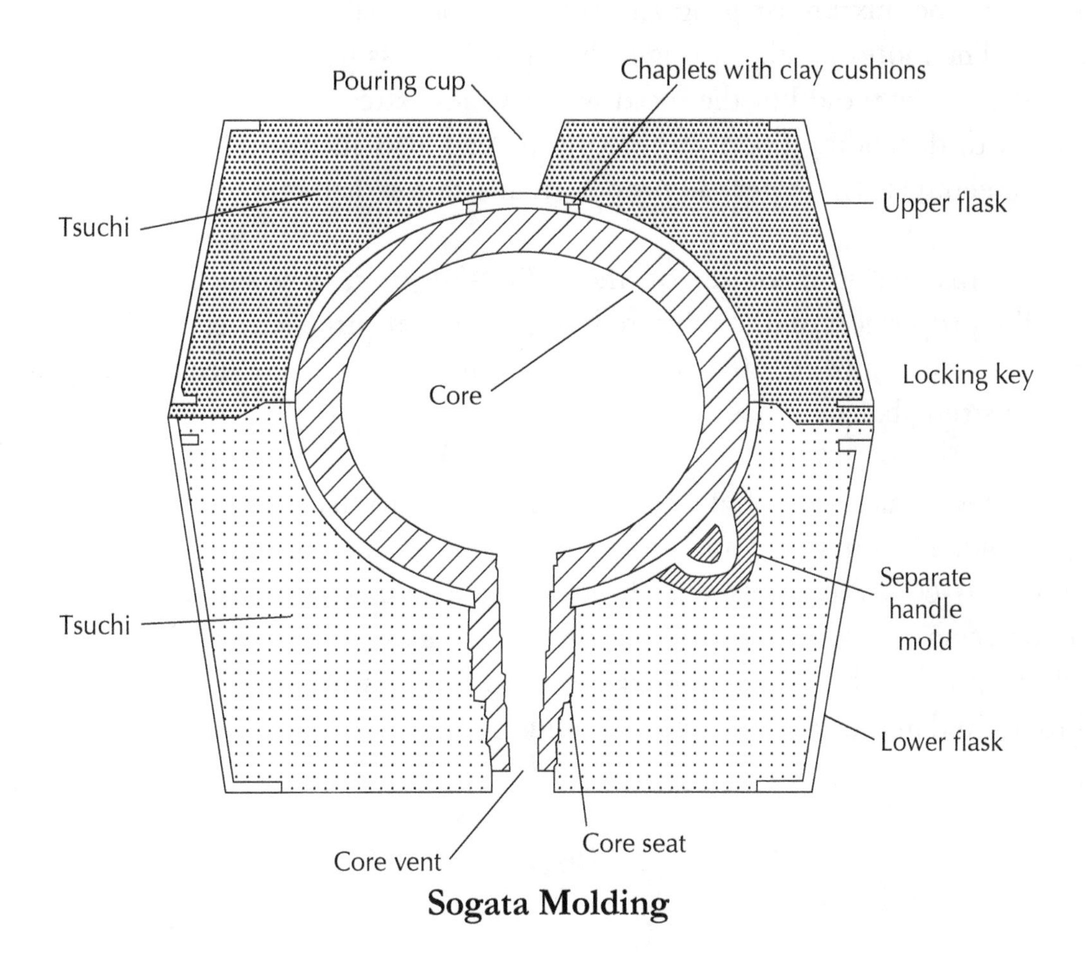

Sogata Molding

Sogata Molding

Sogata molds are made with "Tsuchi," a material made of grog mixed with slip. The grog is made by mixing 90% silica sand with 10% clay (low shrinkage and vitrification qualities), forming this mixture into thin bricks and then bisque firing the bricks to about 1480° F [800° C] . These bricks are then crushed and the resultant grog screened to produce four different grades: a coarse mixture using a #8 through a #15 screen, the next mixture ranges from #20–#30, followed by a #60–#80 screening, with the final grade being a #150 screening. These four grades of the prepared grog are then mixed with a slip made from the same clay used to prepare the grog and water. The mixture of grog and slip must be a balance between strength and porosity. The more slip, the stronger the mold but less porosity, the more grog, the more porous the mold but the mold will have less strength. Between 10–15% clay slip is used, depending upon the grog. One can also successfully use commercial grogs, screening them to get the proper grain size for each mix.

The core material is made up of the #20–#30 grog, new silica sand, and slip. Again the proportion of the mixture is a question of strength versus elasticity and porosity. A core mix of 60% sand, 30% grog, and 10% slip seems adequate to maintain strength.

Separate spout, lug, or knob molds are embedded into the sides or top of the main swept mold. The same clay and grog mixtures are used with different proportions: 50% grog (#60–#80) is mixed with 50% clay. These separate molds are made by hand, pressing a moldable consistency of the tsuchi over the pattern. These two part molds are fired first and then joined in—the case of molds—with cores, or joined and then fired in the case of uncored molds.

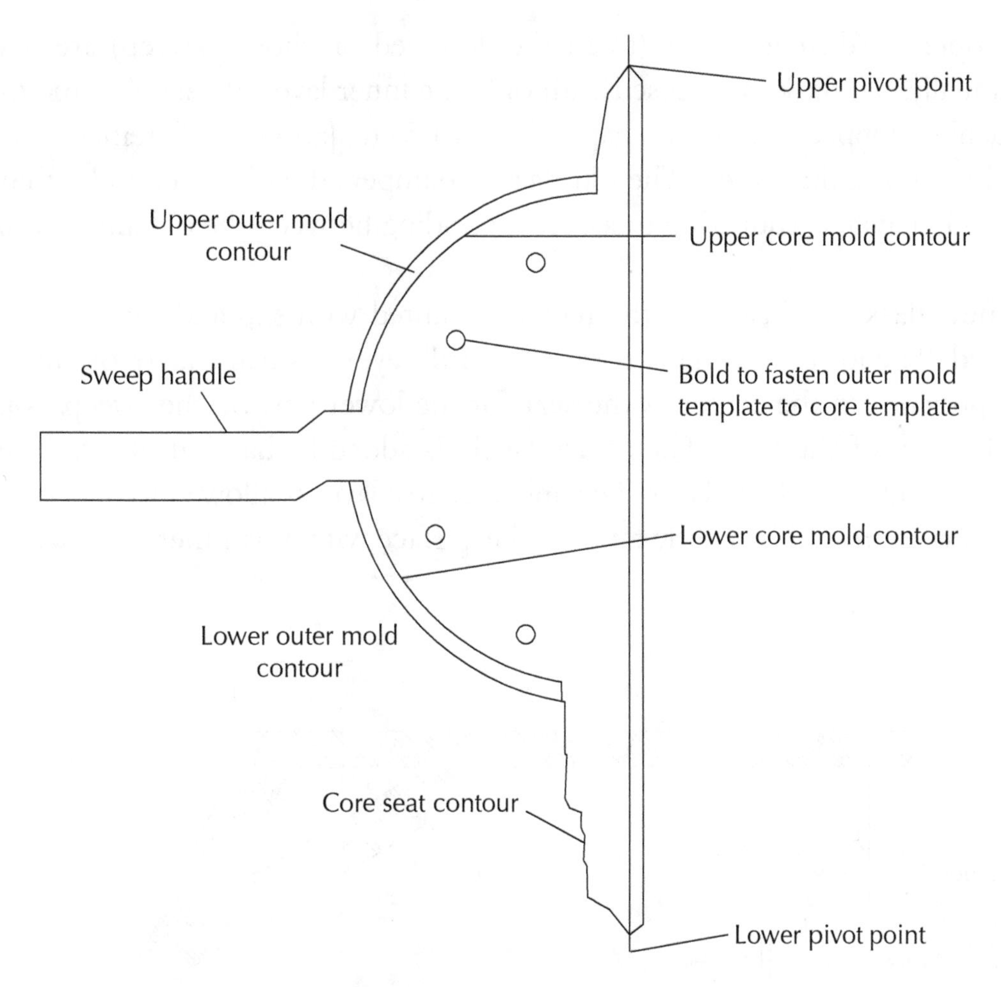

Sweep for Sogata Molding

Traditionally the sweeps to make the molds were fashioned from wood, but today they are made from sheet metal. The main part of the sweep which makes the mold to build the core is constructed from one piece of sheet metal (14–16 gauge) with the metal which will form the upper and lower outside mold contours added to the main body of the sweep with bolts and nuts.

The proper sized metal flasks (to fit the designed or chosen sweep) are selected. Traditionally, old flasks are reused with only the inner layer of tsuchi removed; the old tsuchi is chipped away until a gap of about 1.5 cm. [about 1/2"] remains between the old layer and the sweep. The old flask is dampened with water and a thin layer of the #30 tsuchi is rubbed in to assure a bonding between the old and new layers.

For a new flask, the inside of the metal is painted with slip and the coarse tsuchi is applied. The lower bearing ("Torime"—bird's eye), a small square of metal with a hole punched in the center as the seat for the lower part of the sweep, is placed at the bottom of the flask. The coarse tsuchi is added by hand, using the sweep to check the thickness of the layer. This initial coarse layer is allowed to dry somewhat and then compressed by lightly tapping the surface with a hammer.

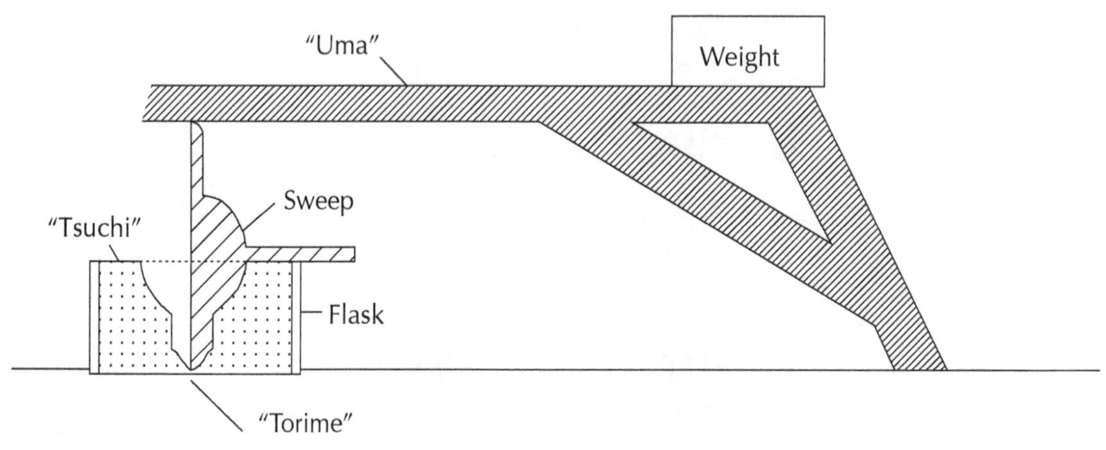

Set-up for Sogata Molding

With the sweep or template in place on the lower bearing and held by the upper bearing "uma"—(horse), the coarse tsuchi is applied to fill in the gap caused by the hammering. The template is removed and cleaned with water and the mold is again left to partially dry. It is important to carefully remove the sweep each time, also being careful to keep the mold in the same position so that the sweep may once again be replaced in an exact position. The next layer of less coarse tsuchi is applied after the surface of the mold has been dampened with a bit of water and slip. The subsequent layers of finer and finer tsuchi are applied in a more liquid form since these layers are filling in the spaces between the coarse grains of the outer layer to produce a fine inner surface. This procedure is repeated until the final application of the very fine #150 mesh tsuchi is applied.

When the finished fine inner surface is complete and is firm and has lost its excess moisture, the surface design may be added. There are three traditional ways of working the surface: modeling tools are used to press various designs into the inner surface, stamps (made of metal or wood) are pressed into the fine tsuchi layer (these two methods produce a raised design), or sand and slip mixtures are added by various methods to the final surface to produce a negative or eaten away design. Very complicated designs are drawn first on thin rice paper ("washi") and the paper pasted on to the inside surface. The design is then transferred to the mold by pressing directly on the paper.

Separate molds for lugs, knobs, or spouts are embedded into the mold surface prior to any design or texture work being done. The mold is carved away at the desired point deep enough to accept the already fired implant mold. The smaller mold is set in place and the surface is filled in and smoothed with the appropriate mesh tsuchi. The pouring cup and sprue are cut into the cope half of the mold, when the molds are dry to the leatherhard stage.

Cores can be made three ways:

1. By packing the core material into the fired mold directly to a thickness of 2 cm., dried, joined with tsuchi, and then shaved to achieve the correct thickness;
2. The mold halves can be lined with rolled out clay having a thickness of the desired metal thickness (part the mold surface and the lined surface with talc before packing in the core tsuchi to the 2 cm. depth), dried and then joined with tsuchi for mortar;
3. Remove the inner mold patterns from the sweep and sweep two new molds, which when packed with the core tsuchi will give the correct sized core, dry, and join as per the other methods. (This method is especially useful when making more than one object of the same size.)

When the mold halves are completely dry, the inside surface is "skin-fired" either by packing them with charcoal and lighting the charcoal or by setting the mold halves upside down over a charcoal forge or fire. The surface is fired to a cherry red color and then allowed to cool slowly.

Cores are traditionally washed with a solution of powdered charcoal, slip, and water; use of the graphite and zircon wash will also work. Mold halves are traditionally smoked over a kerosene smudge pot so that a layer of soot is deposited on the surface; this layer of soot acts as a mold wash.

To assemble the mold, the completely dried and fired drag mold is placed on the floor and the core is inserted into its core print or seat. Small pieces of iron (slightly thinner than the thickness of the casting) called "katamochi" (chaplets) are placed on top of the core; the chaplets are set on a small cushion of damp day. The chaplets make contact with the cope half of the mold and prevent the core from rising when the metal is poured into the mold. The cope or upper half of the mold is then set in place.

The mold is set up on bricks so that the core vent is open to the air. With small molds, boards are placed on either side of the top half of the mold and the pouring team stands on the boards when pouring to keep the cope from floating. Large molds are clamped together. The molds are poured quickly and after a few seconds, the boards are removed and the mold is tipped to spill out the excess metal in the cup. A metal rod the size of the sprue is then inserted to the correct depth to eliminate excess metal in the sprue, eliminating the need to grind off the sprue later!

The traditional method for finishing Sogata castings in iron involves removing all the vestiges of the mold material and then heating the piece in a charcoal fire until it is red hot. The mouth is trued with a iron tapered ring while the piece is still red hot. This process produces a rust resistant iron oxide surface. If a jet black color is desired, the piece is heated again on a charcoal fire and Japanese lacquer ("urushi") is brushed on and is baked into the surface. A beautiful red-brown patina called "Ohaguro" is produced by putting iron in sake (see *Appendix F* for the formula).

Mane-gata

The mane-gata method is used similarly to a sand piece mold in Western casting methodology. The first step in this process involves the production of "tsuchi" similar to the material used in the Sogata process. Two parts silica sand and one part clay are mixed together with water, kneaded to make a moldable consistency, formed into thin bricks, and then fired to 1480° F [800° C]. The bricks are then crushed to make grog and screened into three sizes to make the three different tsuchi.

Again, one can successfully use commercial grogs, screening them to get the proper grain size for each mix.

1. **"Kami-tsuchi" (paper-clay):**
   8 parts #150–#200 mesh grog,
   1 part earthenware clay,
   ten sheets Japanese paper ("washi"),
   water to form moldable consistency.
   Knead or beat to break up the paper fibers (traditionally horse manure is also added to this mixture).

2. **"Tama-tsuchi":**
   9 parts #20–#30 mesh grog,
   1 part earthenware clay slip (7/8 parts water and 1 part clay)

3. **"Ara-tsuchi":**
   8.5 parts #8–#15 mesh grog,
   1 part earthenware clay slip,
   .5 part wood chips, wood flour [sawdust], or chopped straw.

To mold with the Mane-gata process, false piece the pattern in a mixture of green sand and water. Apply light machine oil to the pattern. Cover the exposed pattern with the paper-clay ("Kami-tsuchi") to a depth of 3–5 mm [.10–.20"] . Allow the paper-clay to dry to a leatherhard stage. Apply clay slip (1 part clay and 6 parts water) to the paper-clay layer and add a layer of tami-tsuchi to a depth of 1 cm. [1/2"]. Place bisque fired shards or soft brick pieces on the top of the clay surface to facilitate faster drying.

When the tami-tsuchi layer is leatherhard, remove the shards, and place thick core wire or 3/8–1/2" [.3–.5 cm.] re-bar rods over the surface to reinforce the mold. The thicker re-bar may stick out of the edge of the mold to facilitate tying the mold halves together, but the wire should stay within the boundary of the mold edge. Clay slip is then applied over the tami-tsuchi and a layer of Ara-tsuchi is applied to a depth of 2–3 cm. [1–1.5"]. Shards or soft brick are again used to facilitate the drying process.

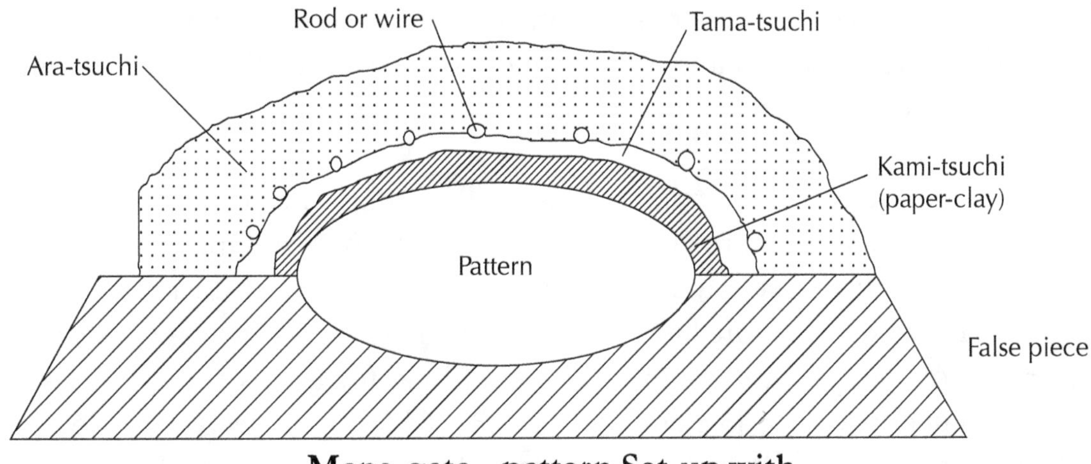

Mane-gata - pattern Set-up with False Piece to Make One Half

False Piece

When the ara-tsuchi is leatherhard, the completed mold half can be rolled over, making sure that the pattern does not slip out of place. Brush away any remaining sand from the false piece, cut several keys in the joint surface, and smooth out the surface. Oil the exposed half of the pattern and part the joint edge of the mold. Now complete the second half of the mold using the same layers of tsuchi as the first half.

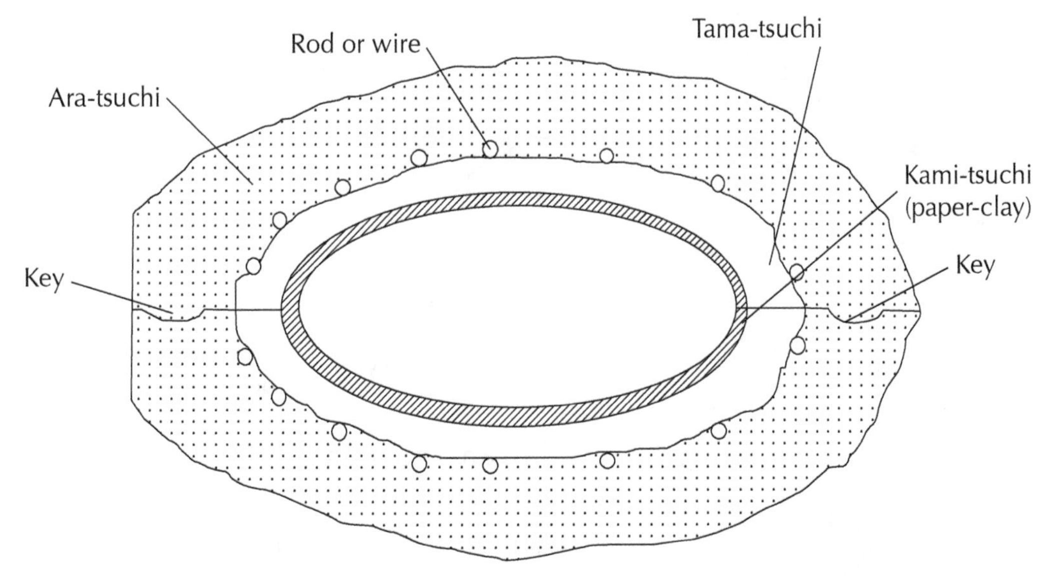

Mane-gata Second Half Completed

When the second half is leatherhard, trim the edges of the completed mold and score marks through the part line for repositioning later. Open the mold and remove the pattern. Reset the mold halves and allow it to dry thoroughly.

Prepare the core material by mixing 1 part silica sand, one part #20–#30 mesh grog, and a thick clay mixture (1 part clay and 1 part water). Open the mold and line the molds with spacer clay rolled out to the desired metal thickness; part the mold interface with talc before lining with the clay. Prepare 3–4 pieces of 3/16" [4.7 cm.] bronze or steel rod cut about 3" [7.5 cm.] long for use as core pins. Using a sharp knife cut a space 1.5" [3.8 cm.] long and 3/16" [.48 cm.] deep at 3–4 locations on the part line surface. Pack in the core to a thickness of 2–3 cm. leaving a vent hole near the sprue location. Cut the channel through the part line for the vent hole and use a small metal pipe to connect the core hollow through the mold to the outside near the sprue. Place the core pins in position. Now make the other half of the core. When the core has dried to near leatherhard, paint a layer of clay slip on the two core surfaces and then place the cope half on the drag half to glue the cores together.

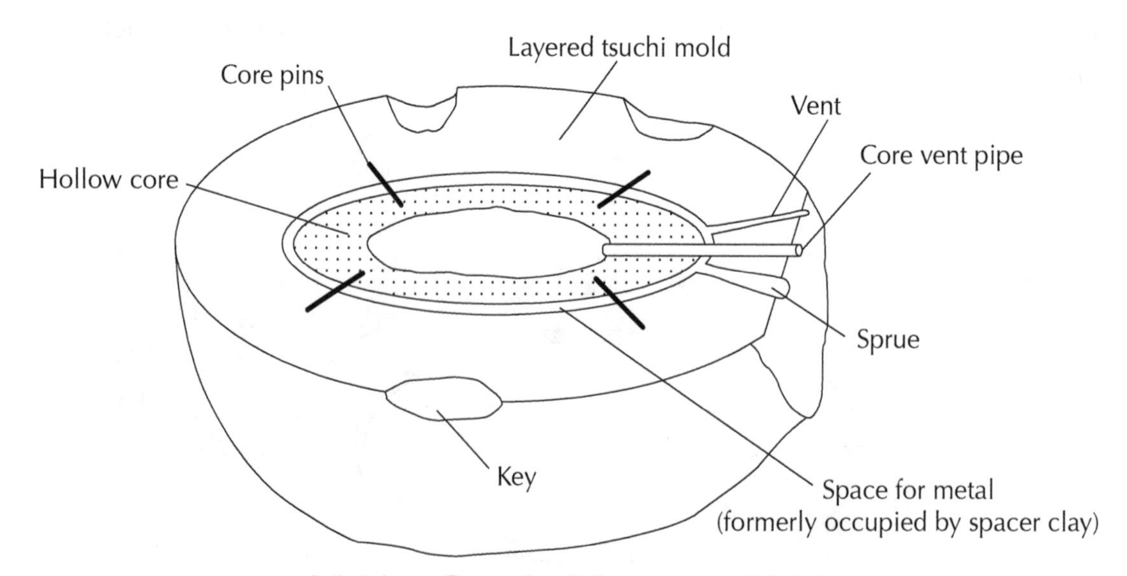

Making Core for Mane-gata Mold

Wait for the slip to dry then carefully lift the cope half of the mold off of the core. Carefully remove the core and take off the spacer clay. Dry the core thoroughly. Cut the sprues and vents and repair any cracks or defects with slip. Place the core in position and fill any gaps around the core pins and core vent pipe with kami-tsuchi. Now place the cope half on the drag half and securely wire together with core wire. Cover the seam and wire with ara-tsuchi. Cores with core prints can also be made, but may also need core pins to help support the core.

The mold is now fired to a temperature of 1480° F [800° C]. Traditional molds are fired in small kilns built to the size of the mold and fired with pine wood. A small mold takes about 3 hours to fire in the traditional way. When the correct temperature is reached, the smoke holes, vents, and fuel openings to the kiln are sealed up and the mold is allowed to stand for several hours. These molds are best poured hot at a temperature of about 500° F [300° C].

Japanese Paper-Clay Molds

Japanese paper-clay is traditionally used in Mane-gata molds and in molding insert molds for inclusion in Sogata molds. One can successfully use paper-clay around wax patterns, curing the molds in the same manner as African molds. Various paper fibers have been tried: Japanese rice paper, Japanese Kozo, Thai Kozo, Abaca, and even newspaper. The paper fiber should be beaten until the fibers are short and well broken up. The paper fiber burns out during the curing of the mold making a very porous mold.

1 part paper (wet)
1 part clay (Red Art or bond)
3 parts fine grog
Water to moldable consistency

Mix to a soft moldable consistency. Apply in the same manner and in the same thickness as in the Bifor method. The pour cup should be an open cup rather than an enclosed cup; this material is not suitable for the inversion casting method. Fire in a simple kiln fueled with charcoal until the molds are dull red. Pour these molds hot.

Evaporative Pattern Casting Process (EPC)

The EPC process is a technique that uses patterns made from expandable polystyrene (EPS), compacted in loose silica sand, and replaced by the metal that is poured directly into the pattern.

Patterns must be made from low density expandable polystyrene (density of 1.0 lb./ cu. ft. to 1.51b./ cu. ft.) in either the bead or foam texture. EPS comes in a wide variety of shapes, sheets, and forms; most packing materials (peanuts, shells, beads, etc.) and insulating sheets are EPS. **Avoid using fire resistant EPS sheets sold at building supply stores (often they are blue in color): likewise avoid polyurethane, which will not burn out (often yellow or tan in color).** EPS can be cut with sharp knives, cut on band saws, sanded, and glued together to form the sculptural pattern. Blehm Styrene Adhesive BP 8002 (http://blehmplastics.com) is the recommended adhesive. (Apply in thin layers like rubber cement and press together.) Other adhesives such as white glue and rubber cement work, but have tendency to retard the flow of the metal. Use straight pins to hold pieces together while gluing. Alter the surface of the foam by using Blehm Pattern Sealer BP 8005 (a sort of soft wax), hot knives, and gluing on thin layers of aluminum foil or paper. Applying enamel spray paint or " magic marker" will bite into the foam dissolving it; by the use of masks or wax resists, a variety of textures or drawings can be achieved. Steel or copper wire, pins, nails, or other thin pieces of metal can be embedded into the foam to also alter the surface texture or the formal qualities of the pattern. EPS less than 1/8" [.3 cm.] in thickness will not cast unless attached to a thicker part.

Remember that the metal will reproduce exactly the form and texture of your pattern.

To set up the pattern for casting, attach a 1" x 1" [2.5 x 2.5 cms.] EPS sprue to the pattern. Top pours work best, however some patterns might be best poured with a bottom pour configuration. Multiple patterns should be attached in a star-like configuration around the sprue. The sprue should extend 3" above the pattern. A EPS "Styrofoam" cup can be glued on the top of the sprue, or a self-set sand cup or small clay flower pot may be used when the piece is embedded in the dry silica sand.

Weigh the pattern and convert weight of EPS to metal:

1 oz. [.03 kilos] EPS= 6.6# [3 kgs] aluminum
22.25# [10.1 kgs] bronze
18.75# [8.5 kgs] iron

Add the weight of the cup. A cup the size of a regular Styrofoam coffee cup will weigh:

.66# [.3 kgs] aluminum
2# [.9 kgs] bronze
2# [.9 kgs] iron
Record total weight.

Note: # = pounds

When all the glued parts are dry, spray or brush the entire pattern with pattern facing. (By volume: 50% zircon, 50% graphite, isopropyl alcohol to sprayable consistency.) Correct facing thickness is approximately 1/32" [.08 cm.] or until small scale surface detail is no longer visible. The pattern facing should be thick enough to capture detail yet thin enough to permit gasses to escape into the sand. Blehm also makes special mold facings for lost-foam casting.

Select 5 gallon metal bucket or other appropriate metal container for pattern to be poured. Allow 3" [7.5 cm.] between the bottom of pattern and the bottom of the container and 2" [5 cm.] between the pattern and the wall of the container. Add the height of the cup above the sprue. Place container on a vibrator platform. Pour 3" of #30–50 AFS screened silica sand (no fines) into the container. Bagged sand without many fines, such as from Badger or Unimin is recommended.

Place pattern in the center of container and carefully pour sand around pattern. **Wear dust mask when handling, pouring, and compacting silica sand!** Sand should be added slowly; take care that hollows and cores are completely filled (patterns with cores should be positioned so that core openings are at the top). Fill sand to within 1/2" [1.3 cm.] of the top of sprue and position sand or flowerpot cup over sprue (or fill completely if using Styrofoam cup). Add remaining sand and then vibrate the container to compact sand. Add additional sand following initial compaction (if needed) and vibrate again. Sand should be level with top of cup and the top of the container.

The sand can also be compacted with a handheld vibrator (such as holding an oscillating sander or an air-powered chipping hammer against the side of the barrel), by rocking the barrel back and forth over a steel welding rod, or by gently tapping with a hammer. Obviously, the vibrating table works faster and more efficiently, but the other methods will work if a vibrating table is not available.

Set the container in the pouring line and place a casting weight (weights with a "U" configuration work best) on top of cup, making sure that weight is also supported by the edge of the metal container and that the cup is clear for pouring.

The pattern may be poured immediately following the compaction process. During the pouring process, you must keep the cup completely filled. Since the EPS is approximately 2% material and 98% air, the hot metal vaporizes the foam and the pressure of the full cup on the system forces the gasses out through the porous sand. This pressure holds the sand in position, as the metal replaces the foam, and duplicates the pattern exactly. Failure to keep the cup filled will result in a loss of pressure and collapse of the sand, causing loss of detail and form.

This process produces toxic fumes when the metal hits the foam. Wear respirators rated for fumes when casting or only pour these molds outside. Pour teams should be aware of wind direction if pouring outside.

After the casting is cool, pour out the sand and remove the casting. Allow the sand to cool. Shovel the sand into metal sand buckets and mark so that you know that the sand had been used. The sand may be used up to 3 times before it becomes coated with the EPS resin enough to reduce permeability. After three uses, the sand may be reused in cores and other self-set sand molds.

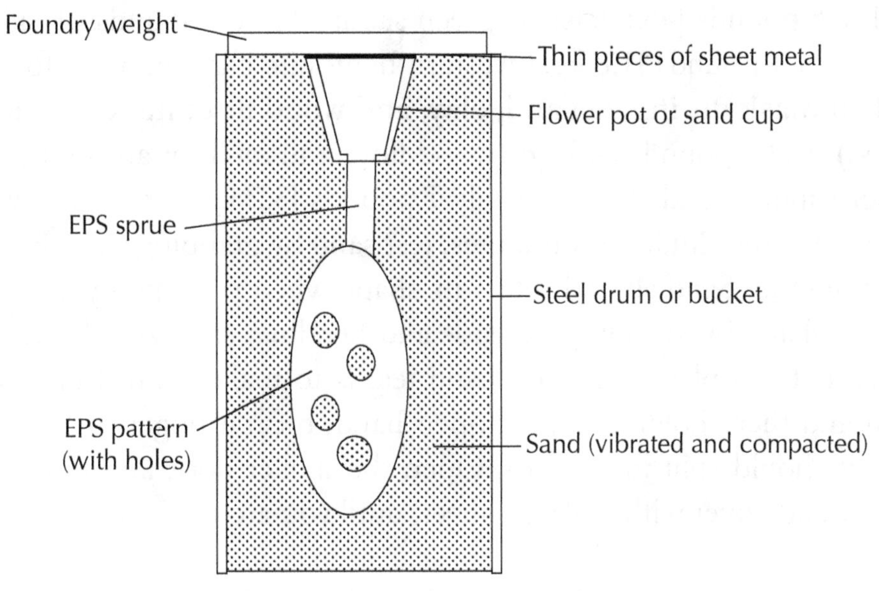

EPC Flask and Pattern Set-up

Green Sand Molding

Green sand molding is the predominate method used in the commercial production of cast metal objects. "Green sand" refers to the fact that the sand remains uncured or "green" and is capable of being recycled. Historically, many molding sands were naturally occurring sands containing naturally occurring clay, such as "Albany sand." Most green sands are "synthetic" sand, in that it is not natural and is "made up." A typical formula is 80–100 mesh silica sand, 4% western bentonite clay (additional clay can be added depending upon the metal being poured, and the type of pattern being pulled), 7% water, and 1% wood flour. As in any molding system, the sand is the refractory, the clay and water are the binder, and the wood flour increases permeability and compressibility of the sand. Additional ingredients, such as seacoal or red iron oxide, can be added depending upon the metal being poured and the type of pattern; limit amounts of additions to 1% or less (you may have to increase clay when adding seacoal or red iron oxide). Store green sand in a metal bin for reuse. Be sure to riddle the used green sand into the bin following its use.

To make 100+ pounds [45.5 kgs] of green sand: add 100 pounds of silica sand to the muller. Add 4 pounds [1.8 kgs] western bentonite clay and mull for 2 minutes. **(Wear a dust mask during all sand, clay, and wood flour handling and mixing operations.)** Add 1 pounds [.45 kgs] wood flour and mull for an additional minute. With muller running, add approximately 7 pounds [3.2 kgs] of water by sprinkling it on the mixing sand. Mull for 4 minutes; test sand by grabbing a small handful and squeezing the sand. Sand should retain its shape when you open your hand. Break the squeezed shape by grasping the ends and bending; sand should fracture along a clean, well-defined plane. Another hand test is to make a small flat patty of the green sand and then, holding it your open hand, push a finger into the center of the patty—it should split into thirds. Open the muller door and mull sand into a wheelbarrow and cover with a plastic sheet, until you are ready to ram the sand.

Green sand castings are traditionally made in wooden snap flasks with metal pin register guides. Patterns are fixed to pattern boards that fit and register with the snap flasks. Patterns should be sealed with lacquer prior to molding. Patterns can be one-sided or split (one half on each side of the board). Patterns can also be free molded or "loose molded" without a pattern board by carefully determining the part line of the pattern and "false piecing" the pattern (embedding the pattern up to the part line and slicking the sand around it), ramming one half, rolling the mold over, removing the false piece and ramming the second half of the mold. If you are reproducing this pattern more than once, it is a good idea to attach a wooden runner and gating system directly to the pattern board as this saves cutting the runners and gates later. Snap flasks should not to be used for any other molding except green sand or Petro Bond sand molding. Resin bonded or sodium silicate bonded molds will stick to the flasks. Molds should not be poured in the snap flasks, as it risks burning or charring the flasks.

To make a green sand mold, set up the snap flasks with the pattern board in between the flasks and a fitted bottom board underneath; be sure the snaps are on the same side or you will not be able to open the flasks! Set the flasks so that the drag (bottom half of the mold) is on top. Part the pattern with talc and set a riddle with 1/4"–1/2" openings [6 mm–12 mm] on the flask, shovel prepared green sand into the riddle, and shake riddle to deposit 1/2" [12 mm] of sand over the entire pattern and pattern board. Tightly ram sand around the pattern and then add another 2–3" [5–7.5 cm.]

of green sand evenly over the pattern and the pattern board. Ram sand firmly with your hands followed by a handheld wooden rammer. Pay particular attention to the edges of the mold and the mold corners. The trick is to ram hard enough to make a solid compact mold that will not drop out of the flask and yet loose enough to make a permeable mold (able to absorb gasses from the casting process). Continue to add sand until the flask is full and then strike off the top of the mold so that it is flat and even.

Place another bottom board on top of the mold and roll it over. If it is a large mold, use metal dogs and wedges to clamp bottom boards and flasks securely together during the roll over. Remove the dogs and the bottom board; the cope (upper half of the mold) is now up. If the pattern is a loose or free-molded pattern, remove the false piece material and gently blow off the pattern with an air gun or brush the false piece material away. Repeat the molding process for the cope half. Be sure the cope half is well rammed and stuck off flat. You are now ready to remove the pattern and the pattern board.

Gently rap the pattern board that is sandwiched in between the cope and drag. Grasp the cope part of the snap flask near the draw and guide pins and pull straight up. The sand should pull free from the pattern. If sand sticks or falls out you have not used enough parting compound or rammed the mold hard enough. Now rap the pattern board or the loose pattern and gently pull it straight up from the drag half of the mold. The sand at the edge of the pattern sometimes pulls away or falls in. This problem can be alleviated by paying more attention to the ramming process near the edge of the pattern. Try rapping the pattern or board a bit more vigorously, which slightly compacts the sand away from the pattern. Or, in the case of loose patterns, use a brush to apply water around the edge of the sand adjoining the pattern (the sand has a tendency to dry out against the pattern) prior to the draw of the pattern.

If the runner and gating system is not molded as part of the pattern, cut the well, runner, and gates into the sand using a slick and spoon, or thin metal gate cutter. Locate where the sprue is to be cut in the cope part of the mold. Roll mold so that the top of the cope rests against a bottom board and, using a thin-walled piece of metal tubing, cut the sprue using a twisting motion. Then using a 1/8" [3 mm] rod, make several "whistler" vents in the cope by pushing the rod from the pattern cavity to the top of the cope. Now roll the cope on its side and using your spoon cut a cup at the top of the sprue. Gently blow both the cope and the drag to remove loose sand; check the mold—if the sand is dry, spray with water and repair any parts of the mold with the slick. Replace cope on the drag using the guide pins, making sure that the snaps are on the same side. Remove flasks by releasing both snaps and opening the flasks. Set the mold on the floor; use the correct-sized metal jacket and push the jacket down over the mold. Weight the top.

Green sand molds should be poured within 4–6 hours of molding; cover molds with plastic during very dry periods. Following the pour, shake out the green sand and put it through a riddle to break up lumps, and store for reuse.

Direct carving of green sand is also possible. Use 6–8% clay to make a harder mold. Ram drag half of the mold, part, and then ram the cope half. Pull the cope and carve in either or both halves using spoons, slicks, knives, or by pressing textures and objects into the mold. Cut sprues, runners, gates, and vents. Replace the cope on the drag and set up for casting as any other green sand mold.

Lost-Wax "Melt-in" with Green Sand

Small simple wax patterns can be molded in green sand, the molds placed in an oven, and the wax pattern eliminated. Patterns should be top gated if possible. Find a metal container, such as a coffee can, #10 can, five-gallon bucket, or other suitable container. The metal container should be sized so that there is at least 3" [7.5 cm.] between the bottom and sides of the pattern and the bottom and sides of the container. The wax pattern should be coated with mold wash in the same manner as in the melt-out procedures in chapter 6. Make up the green sand with 6% western bentonite and the minimum amount of water necessary to produce a satisfactory "squeeze and break" test. Add 3" [7.5 cm.] of sand to the bottom of the container, ram firmly, then riddle in 1/2" [1 cm.] of sand and place the wax pattern snugly into the sand. Add riddled sand around the pattern, taking special care to tuck the sand firmly around the wax pattern. Continue to ram the sand up the pattern and up the sprue and vent until the sand is level with the top of the container, remembering to also pack the sand firmly along the sides of the container.

Strike off the top of the mold and fit a resin bonded sand cup over the sprue wax and a sand vent over the vent wax. Retain the cup and vent to glue on the hardened mold following the melt-in procedure. Place the container of sand with the wax pattern inside, top side up into the oven and heat to 350° F [176° C] for 12–24 hours. The wax melts into the sand and the mold hardens as the water is driven off. These molds are porous enough for the wax to travel a long way into the mold and still retain enough permeability to absorb the gasses generated in the pouring process. Remove the hot mold from the oven, glue on the cup and vent sand, and pour while still hot. The sand can be recycled except for the sand that contains the wax residue. This method is an easy and inexpensive way to cast simple wax patterns; it is not particularly effective for complicated wax patterns because of the difficulty of ramming the sand tightly and evenly against such patterns inside a rigid container.

Molding with Petro Bond

Petro Bond is a "green" sand which is bonded with oil instead of water. The materials to make Petro Bond can be ordered from the Smelko Company [http://smelko.com]. Petro Bond may be used to mold the same kind of patterns as regular green sand. The advantage of Petro Bond is that is does not dry out and it is reusable without re-mulling. Petro Bond also has more green strength than most water based green sands, making it easier to mold. It is especially useful for false piecing loose patterns for resin bonded sand molds. Petro Bond can also be mixed with very fine sands for the casting of very detailed reliefs, signs, and pattern letters. It is recommended that you store Petro Bond in its own bin to avoid contaminating it with green sand or self-set sand materials.

Add the sand to the muller and then mull Petro Bond clay into the sand for 2 minutes. Add the oil and mull for 4 minutes. Add the catalyst and mull for an additional 4 minutes. The sand should exhibit the same qualities as regular green sand when using the squeeze and break test. Use the Petro Bond in the same manner as the regular green sand. Be sure to put the Petro Bond through the riddle after its use to break up rammed lumps and make it ready for its next use. Return the Petro Bond to the storage bin. If the Petro Bond begins to lose its green strength, it can be restored by mulling in 1 pound [.45 kgs] of Petro Bond and 1/2 oz.[.05 kilos] of catalyst.

To make 100 pounds [45.5 kgs] of concrete molding sand:

1. Put 100 pounds [45.5 kgs] of silica sand into the muller.
2. Add 10% High-Early Strength Cement [10 pounds or 4.5 kgs], and dry mix.
3. Add 8% or 8 pounds [3.6 kgs] water by sprinkling the water into the sand while the muller is running.

Mull till sand achieves green strength and mold in the same way as green sand or melt-out molds.

To make 100 pounds [45.5 kgs] of Petro Bond:

100 pounds [4.5 kgs] 60–80 mesh silica sand
5 pounds [2.3 kgs] Petro Bond
2 pounds [.9 kgs] oil
1 oz. [.03 kgs] P-1 catalyst

Molding with Concrete Molds

Cement can be used as a binder to make strong molds. Molds for heavy castings in iron and steel are often made with concrete molds. Patterns should be well sealed with lacquer and parted with talc or graphite. Wax patterns can also be melted out of concrete molds by using the same methods as for melt-out molds from self-set sand. Concrete is cheap and easily available making this method ideal in situations where it is difficult to get alkyd oil or sodium silicate binders.

Although the cement will set up the mold, these molds need to be dried at a low temperature. Do not heat high enough to calcine the cement. Melt-out molds should be dried at 300° F [149° C]. Piece molds should be dried at 200° F [93° C].

Melting and Pouring Metal

Chapter 8

Melting and Pouring Bronze and Aluminum

Melting and pouring metal is hot, tiring, yet exciting and satisfying work. Unless you are set up as a one-person shop, it takes teamwork to pour your pieces. Decide on which persons will tend the furnace and who will pour, dead-man, and skim during the pouring operation.

Bronze and aluminum are usually melted in crucible furnaces fired with forced air and natural gas. Propane and other fuels (such as coke and charcoal) can also be used; some fuels require furnace modifications. Induction furnaces will also melt nonferrous metals. Crucibles are made from clay and graphite or from silicon carbide. While clay/graphite crucibles are less expensive, silicon carbide crucibles are stronger and last longer. Furnaces are sized to fit the size of the crucible. Smaller crucibles can be fired in larger furnaces, but this practice is less efficient. Crucibles are numbered according to the number of pounds of melted aluminum that they will hold. A #30 crucible will hold 30 pounds [13.6 kgs] of molten aluminum. Since bronze is approximately 3 times as dense as aluminum (aluminum = .093 pounds/ cu. inch; bronze= .309 pounds/ cu. inch), multiply by 3 to get the number of pounds of molten bronze a crucible will hold. A #30 crucible will hold 90 pounds [40.1 kgs] of bronze.

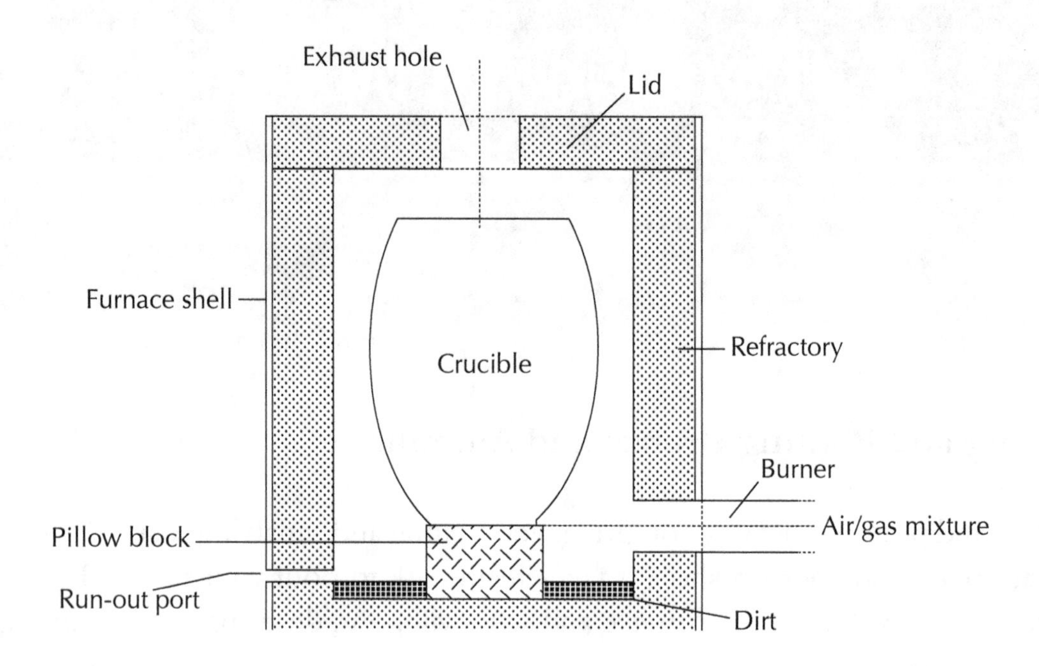

The furnace is a steel shell lined with refractory. The crucible sits on a pillow block so that the flame hits right at the base of the crucible and tangent to its exterior wall. Simple burners can be constructed using vacuum cleaner blowers, steel pipe, and a gas valve. Most commercially made melting furnaces, such as "Speedy Melt "made by McEnglevan (www.mifco.com/hsmelter.htm) have extensive electronic controls and usually need to be professionally installed.

Gas Furnace Set-up Procedure Remove the furnace lid and clean spills and dross from the furnace bottom. Remove the old dirt layer. Make sure the pillow block is centered in the bottom of the furnace and then ram 1" [2.5 cm.] of damp dirt around the pillow block. The dirt level should be below the burner port. Wash the sides of the furnace, underside of the lid, the bottom of the crucible, and the top of the pillow block with furnace wash (graphite, zircon, and water to a paintable consistency). Washing the bottom of the crucible with mold wash will also help keep the crucible from sticking to the pillow block.

Select the proper size crucible for the furnace and the alloy being melted. Do not melt aluminum in a bronze crucible or bronze in an aluminum crucible! Check the crucible for chips, cracks, or holes. Be sure the pouring lip is free from dross or frozen metal. Check to be sure that the crucible fits in the proper pouring shank and that the pick-up tongs fit properly. Some pouring shanks require lugs to fit the crucible snugly. Usually three lugs are required; position the lugs so that two are to the pouring lip side of the shank and one is near the keeper hook side. Check to see that the keeper hook fits and that it can be tightened down.

Place two squares of cardboard on top of the pillow block. The cardboard will form a thin layer of ash which will help prevent the crucible from sticking to the pillow dock when it is hot. Then place the crucible in the furnace, centering it with the lip facing away from the burner port and to the front. Weigh out the proper amount of metal for the crucible size being used. Scrap should be 30–40% of the total weight. Charge the crucible with clean scrap from the proper scrap bin. Be careful not to wedge the scrap tightly into the crucible, or the metal will crack the crucible as it expands with the heat. Be sure that the scrap is not piled up over the top of the crucible.

Spread a layer of damp dirt from the pit around the furnace and in front of it. Place three clean bricks in a triangle near the furnace as supports for the hot lid when the lid is removed (Some furnaces have swing away lids, that swing to the side of the furnace.) Setting the hot lid directly on the concrete may cause the concrete to explode. Set up a skimming grate near rear of the furnace and assemble skimmers, the two pouring shank rests, the keeper hook, tongs, and ingot molds. Set the pouring shank on the floor in front of the furnace so that the keeper hook holder is toward the rear of the furnace. Set the pouring shank up on bricks so that it can be easily picked up while wearing gloves. Be sure that your set-up of the lid, skimming grates, and other tools are out of the way of pulling the crucible out of the furnace.

After the furnace has ignited, center the lid on the furnace. Never start the furnace with the lid on the furnace. Be sure all your ventilation equipment is on and functioning. Place the remaining scrap charge and ingots on the lid to warm. Do not block the exhaust hole in the lid with scrap or ingots.

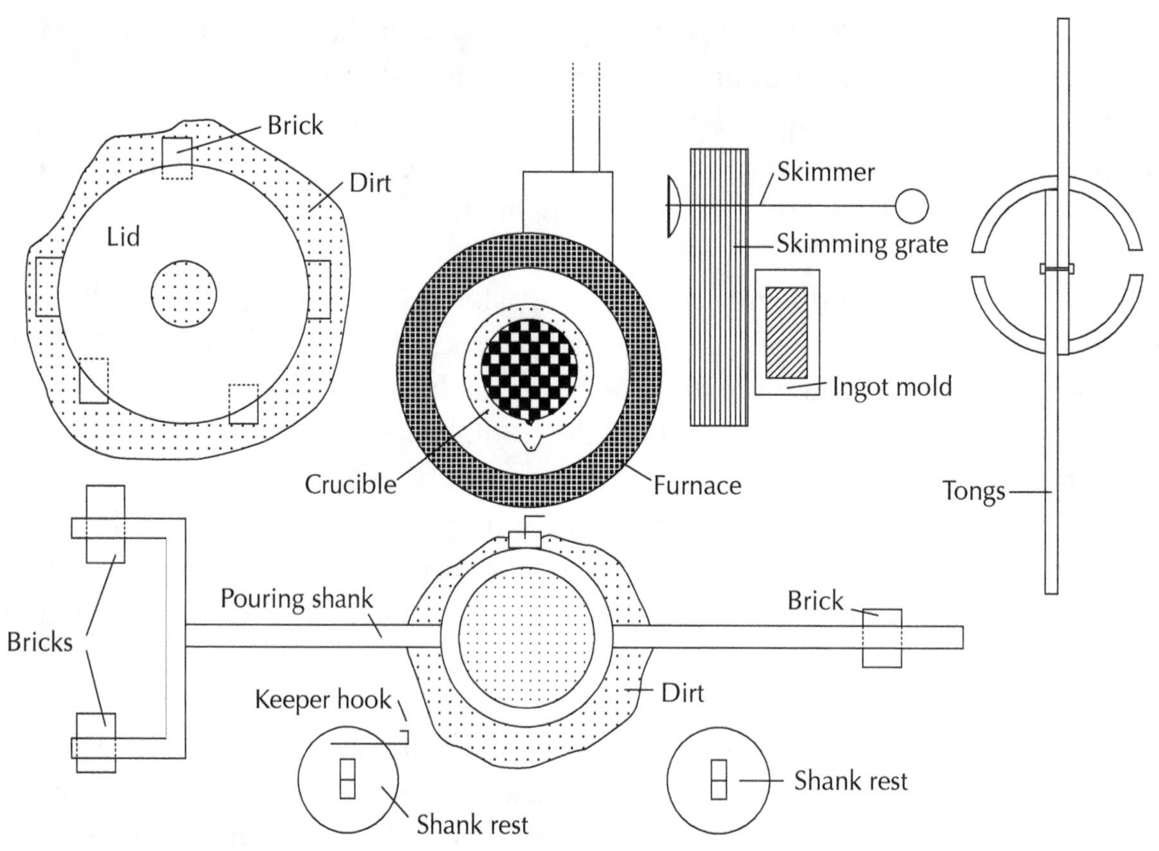

Top View: Furnace Set-up

Melting Procedure Always melt clean, oil-free metal. Do not mix alloys and do not melt "mystery metal." Scrap metal (sprues, vents, cups, and spills) should be melted with new ingots. The continual remelting of scrap without adding new metal results in "short metal" as the alloying metals becomes depleted. **Always preheat any scrap or ingots before charging them into a hot crucible of molten metal**. Scrap and ingots absorb moisture that can cause an explosion of molten metal and cause absorption of gas (hydrogen) into the metal. Small pieces of metal can be placed in the crucible with a pair of tongs. Turn the furnace down and remove the lid to charge ingots or large pieces of scrap. Do not drop ingots or large pieces of scrap into the crucible or jam them against the sides of the crucible.

Charge the metal and melt as quickly as possible. This requires constant tending of the furnace. When the melt is near the top of the crucible, charge only small pieces of scrap so as to not over-fill the crucible. Do not overheat the metal or hold the metal in the crucible for long periods of time.

Most bronze alloys melt at 1750° F [954° C] and have reached their pouring temperature when you can observe the following signs in the crucible: The dross is melted, shiny, and whipping up in a small cone toward the center of the crucible; the surface of the metal is shiny and mirror-like; and eddies of metal can be observed flowing up from the edges of the crucible and into the center, or there is a distinguishable eddy formed in the center of the crucible.

At 2200° F [1200° C], the eddies form a bright circular form in the center of the crucible known as the "Eye of God." Of course, measuring the temperature with a digital pyrometer will assure exactly the proper temperature for pouring. One could make a simple pyrometer by wiring a metal shielded chromal/alumal pyrometer tip (often available as a replacement part) to a millivolt meter. Millivolts can then be converted to degrees Fahrenheit or Centigrade by consulting a conversion table (One is available at http://www.thermibel.be/documents/thermocouples/thermocouple-tables.xml?lang=en).

For aluminum, the metal has reached the proper pouring temperature when the surface of the metal changes from silver to pink. Most aluminum alloys melt at 1220° F [676° C], and thin castings are poured around 1350° F [732° C]. "Coverall" made by Foseco (http://www.foseco.com/en-us/end-markets/foundry/products-services/non-ferrous-foundry/non-ferrous-foundry-detail/productsinfo) makes a good flux for aluminum as well as grain refiners and degassers. (See *Appendix G* for fluxing and degassing procedures.)

The furnace should fire with a slight oxidation flame (a green-yellow flame observed from the exhaust port in the lid). An oxidizing flame reduces the amount of hydrogen that is absorbed into the molten metal.

Pouring Procedure The molds should be set up on the floor with 1" [1.25 cm.] of dirt underneath or in the pit and grouped according to the thickness of the casting. Be sure that the pit sand has been wet down prior to the pour. "Thin" molds (castings of 1/8–1/4" [3–6 mm thick]) should be at the front of the mold line or grouped together in the pit, "medium" molds (castings of 1/4–1/2" [6–12 mm thick]) next, and "thick" molds (castings of 1/2" or over) last in line.

For bronze, thin castings are poured first when the metal temperature is 2150–2050° F (1177–1121° C), medium castings are poured between 2050-1950° F (1121–1066° C), and thick castings are poured at 1950-1850° F (1066–1010° C). This temperature information applies to 12-A silicon bronze. For the ranges for other alloy temperatures, consult: "Copper-based Alloys Foundry Practice," an American Foundry Society (AFS; www.afsinc.org/) publication.

For aluminum, thin castings are poured at 1400–1350° F [760–732° C], medium castings at 1350–1300° F [732–704° C], and thick castings are poured at 1300–1250° F [704–676° C] Aluminum seems to be a most "forgiving" metal and temperature ranges are less critical. A nice aluminum alloy is #319 containing Al 89%, Si 6%, Cu 3.25 %, and a trace 1.75% of Fe. Clean, sand cast scrap is also melted when available. Avoid melting extruded aluminum. Mixing many aluminum alloys can sometimes result in short aluminum with a very large crystal structure that is easily broken.

Molds should be placed on the floor or in the pit so that the pouring cups are easily accessible for pouring; all cups must be at the same height or separated sufficiently to allow access by the crucible and pouring shank. Molds on the floor should be flasked and rammed securely with damp dirt from the pit. Be sure to wet down the pit prior to lighting the furnaces. Spread the dirt about 2 feet [60 cms.] around the pouring area.

Piece molds and roll-over melt-out molds should be banded and weighted. The amount of weight placed on top of the mold should be at least 3 times the weight of the casting. Large flat castings should be weighted with 5 times the weight of the casting or more. Most of the time too much weight does not cause any problems, which is not the case with too little weight. Be sure to position the weights so that the pouring cup is accessible. Very large molds should be secured with pipe clamps.

These molds can be positioned on pallets and the pipe clamps are tightened between the channel iron beneath the pallet and channel iron on the top of the mold.

When the metal has reached the proper temperature and all the molds are in position, the pouring team assembles with the proper safety equipment. An adequate "pouring team" consists of a pour person, the "dead-end" person, the skimmer, and a relief person who stands ready with a shovel to control any run-outs or to provide relief should someone get tired.

Turn the furnace down to its idle position, remove the lid, and place it on the bricks and dirt. Most commercial furnaces will have lids that swing away to access the crucible. Skim the slag or dross from the surface of the crucible and deposit it on the skimming grate. **Be sure that all tools that touch the metal have been heated prior to their use!!!!!)** The pour person and the dead-end person use the pick-up tongs to lift the crucible straight up and out of the furnace. Remember that the tongs are designed so that you operate the opposite side of the tongs; do not set the tongs down on the crucible, as they may crush the top of the crucible when you lift, but make sure the tongs are below the bilge (widest part) of the crucible. Set the crucible down in the pouring shank so that the lip is 90 degrees to the line of the shank's carrying bar. The skimmer replaces the lid and then skims the metal one more time, paying particular attention to the lip area of the crucible. The pouring team now picks up the shank evenly and sets the shank on rests while the skimmer raps the shank on either side of the crucible with the skimming tool to set the crucible in the shank. The skimmer then inserts the keeper hook so that it hooks over the back edge of the crucible (opposite of the lip) and tightens the bolt against the hook, so that the crucible will not fall out of the shank during pouring. Some commercially made shanks have keeps that are activated by the pour person.

The person at the pour end of the shank directs the pouring team. The skimmer helps the pour person line the lip of the crucible up with each cup. When pouring, the lip of the crucible should be as close to the cup as possible. Remember, liquid metal will pour out and away from the lip. It is most difficult to pour when the crucible is full, because there is little lead time before the metal spills out of the crucible. Pour the metal as quickly as possible, keeping the cup filled up. Do not splash, drip, or spill metal into the cup. The skimmer's job is to make sure that no dross flows into the cup; the dross should be gently skimmed back without disturbing the flow of the metal. Do not keep the skimmer in the pot. Observe and skim any dross back if it forms on the metal surface and is in danger of flowing into the mold. Keep pouring until the metal comes up in the vents or risers.

Continue pouring the molds according to the determined order. Be sure to determine how many molds can be poured with each pot before removing the crucible from the furnace; no one wants their mold to be poured short. It is better to have metal left over than not to have enough on the last mold poured. Excess metal is "pigged out" (poured into preheated ingot molds) or left in the crucible if more molds are to be poured.

If another heat is necessary, wet down a cardboard square and place it on the pillow block just before the pouring team replaces the crucible in the hot furnace. If no more molds are to be poured, the skimmer scrapes the inside of the hot crucible clean before it is removed from the shank.

Pouring with the Inductotherm Lift-Swing Induction Furnace

The Inductotherm Lift-Swing furnace holds two #50 clay graphite crucibles, potentially enabling 300 pounds of bronze to be poured in one heat. The crucibles do not get red hot in an induction furnace as the metal is heated by inducing the heat directly in the metal from the coil surrounding the crucible. The lift-swing part of the induction furnace holds the water-cooled coil and can be positioned over one of two crucibles at one time.

When charging the induction furnace, do not put ingots or scrap into the furnace while the power is flowing to the coil!! Wait for the operator to turn down the power before adding the metal. Wear earplugs while working around the induction furnace, as the furnace emits a high frequency sound.

When the metal is melted and is at the correct temperature, assemble the pouring team. Five people are needed: the pour person, the "dead-end" person, the skimmer, the crane operator, and the relief person. Place the bail for the lift-swing crucible shank in the hoist and move the bail and the shank with the crane to the far right of the furnace. The basket part of the shank should now be lined up with the crucible. The operator will operate the lift, raising the coil above the crucible and swinging it into the second position. Open the basket and lift the shank over the crucible to the far left of the crucible. Slip the tongue of the basket under the crucible until the shank is securely under the crucible.

The skimmer now closes the basket around the crucible and secures the pin that keeps the basket from opening during the pour. The skimmer then sets the keeper hook and tightens the screw to hold the keeper hook securely in place during the pour. The skimmer and the relief person move the bail under the shank and the crane operator lifts the crucible off of the pad. Use care in running the crane!! Lift slowly as the crucible is full at this point and jerking the crane will spill the metal. Move the crane into position and line up the crucible with the mold line. The crane operator is responsible for lowering or raising the crucible to accommodate each mold. The pour person is totally in charge and gives directions to the crane operator as to how to position the crane to pour each mold.

Shake-out and Clean-up When the pour is over, everyone involved helps replace equipment in its proper place and helps shovel the dirt back into the pit. The foundry should be totally swept down and cleaned up before anyone involved in the pour leaves. Molds should rest undisturbed for at least one hour following their pour. Molds can be removed using a wheelbarrow, metal 4-wheeled cart, or a 2-wheeled hand truck to an area for shaking out. Use a hammer and strike the mold on a flat face to break it up into large chucks. Be careful not to hit your piece as the metal is still soft and you may break or damaged its surface.

Bronze Alloys

Name	ASTM#	Cu	Sn	Pb	Zn	Al	Si	Fe	Mn
Tin Bronze	1A	88%	10%		2%				
Leaded Tin Bronze	2A	88%	6%	1.5%	4.5%				
Yellow Brass	6A	72%	1%	3%	24%				
Red Brass	4A	85%	5%	5%	%5				
Aluminum Bronze	9A	88%				9%		3%	
Nickel Silver	10A	57%	2%	9%	20%				12%
Silicon Bronze	12A	92%			4%		4%		
Silicon Bronze	13B	95%					4%	trace	trace
Chinese Bronze		80%	4%	10%	2%			4%	
Greek Sculpture		88%	6-9%	trace	trace	trace	trace	trace	
Italian Renaissance		86%	12%	2%					
French Bronze (17th C)		90%	2%	1%	7%				

Metal Theory

As heat from the furnace is transferred to the cold metal in the crucible, its temperature rises until it reaches the range of solidification where it begins to become soft and mushy and then turns liquid. The metal continues to be heated until it reaches 50–100° F [10–37.7° C] above the pouring temperature. Bronzes like silicon bronze are very liquid and can be cast at slightly lower temperatures and in thinner sections. Superheating the metal above its pouring temperature also allows the founder to let the metal stand in the crucible for several minutes after it has been removed from the furnace, facilitating the release of hydrogen gas dissolved in the liquid metal.

After the metal has been poured in the mold, it begins to solidify. Silicon bronze belongs to a group of copper-based alloys with a short freezing range (range of solidification); these alloys freeze by skin formation. That is, after the mold absorbs the super-heat from the metal, solidification begins in the metal adjacent to the mold wall as that metal falls to the freezing point. Freezing occurs as numerous small crystallites form at the mold wall and then rapidly grow sideways and forward into the molten regions of the casting. These crystals link up with their neighbors to form the solidification front, which moves into the interior of the casting perpendicular to the mold wall.

Casting Defects and Their Prevention

1. **Misrun or cold shut-** a casting that is incomplete because the mold cavity has not been completely filled with metal or where the metal has failed to join or coalesce.
   A. increase height of cup and length of sprue (not enough pressure on the system); increase number of gates
   B. metal too cold, pour hotter
   C. cup not kept full during pouring or metal poured too slowly; metal splashed into sprue

2. **Swell-** a deformation on the casting due to sand displacement by the metal pressure.
   A. insufficient weight on the mold
   B. mold not rammed uniformly

3. **Burn-in-** the surface of the casting is rough as the metal penetrates into the spaces between the sand grains, sometimes enclosing or encapsulating the grains.
   A. metal poured too hot
   B. improper amount of binder in the mold causing a soft mold
   C. bad ramming of the mold; sand not packed evenly and tightly enough
   D. core wash applied unevenly or too thinly

4. **Shrinkage-** the casting has holes or depressions on the surface.
   - **A.** pattern designed with adjoining thick to thin sections; pattern should have uniform thickness or a gradual change from thick to thin
   - **B.** insufficient gating system or gating system placed so that directional solidification does not take place (thin sections should be farthest from the cup and sprue; risers over thick sections)
   - **C.** metal poured too hot

5. **Hot tears or cracks-**
   - **A.** metal is "short" (melted too many times without being replenished with ingots)
   - **B.** metal poured too hot
   - **C.** mold too hard (especially true of cores that are too hard)
   - **D.** sharp comers in pattern design or thin sections joined to thick sections
   - **E.** gates and risers located so that they pull against one another as the metal cools; "dumb-bell" design of pattern (two thicks joined by a thin)
   - **F.** casting removed from the mold too soon

6. **Inclusions-** sand particles, mold pieces, or oxides of metal trapped in the casting or on the surface.
   - **A.** improperly mixed sand or mold rammed too loosely (metal washes sand from mold surface as it flows)
   - **B.** sharp comers in the gating system or small projecting mold parts which break off and are washed into the mold cavity by the metal
   - **C.** oxides are formed by poor skimming of the crucible, failure of the skimmer to hold back dross or slag during the pour, metal turbulence during the pour (sprue and gating system should be smooth and even), or metal splashed into the mold

7. **Porosity-** the casting is marked by tiny holes in the surface or through the matrix of the metal (caused by hydrogen gas dissolved in the metal).
   A. poor melting practice: metal containing moisture introduced into melt, reduction flame on furnace, use of cold skimmers, rods, or tongs
   B. metal poured too hot
   C. metal not allowed to stand for a minute or two (to allow gas to escape) prior to pouring
   D. patterns with thick and thin sections tend to have gas in the thick sections
   E. the metal should be poured so the cup stays filled and so that air is not pulled into the mold

Melting and Pouring Other Metals

Lower melting point metals such as lead, tin, zinc, and pewter are easily poured into a variety of molds. Dry plaster molds will accept many of these low temperature metals. Unorthodox molds such as molds made from wood can also be used. Tin and lead can even be poured into silicon rubber molds. The melting point of tin (Sn) is 450° F [232° C], lead (Pb) is 620° F [327° C], and zinc melts at 785° F [419° C].

Pewter is an alloy of tin and lead, tin and antimony (Sb), or tin, antimony, and copper. With tin and lead alloys, the higher the percentage of lead, the softer and grayer the metal; the higher the percentage of tin, the harder and brighter the metal. Pewter containing lead should not be used with any form in which food or drink is stored or served.

Some traditional formulas for lead-free pewter are:

Name	Sn	Sb	Cu
Fine Pewter	82%	18%	
Common Pewter	83%	17%	
Superior Pewter	85%	15%	
Britannia Metal	96.5%	1.5%	1.9%

In alloying any metal, the higher-temperature melting metal is added first to the crucible and melted, then the next highest melting point metal, and so on until all the metals have been added to the crucible. Pewter alloyed with antimony will be less brittle if it is alloyed first, poured into ingots, and then remelted before pouring pieces. Tin and lead can be melted in an aluminum or iron pan on a regular stove. Be sure not to overheat lead and produce toxic fumes. **Ventilation systems should be on when melting and alloying these metals.**

Casting Iron

As a material for cast sculpture, iron is less expensive than bronze, has a very different look and feel, has a very different history, and patinas very differently than bronze. Iron has a relationship to industry and machinery which makes it less of an "art" material, with a more "common" pedigree. Pure iron melts at about 2750° F [1510° C], but most iron alloys have lower melting points. We try to reach a temperature of 2400–2500° F. for most of the molds poured.

Many sculptors casting iron use high phosphorous grey iron (93% Fe, 3% C, 1.5% Si, 1% P, and .5% trace of other elements); radiator iron is used, because it is easily broken up into the required sizes for charging into the cupola or cupolette. For iron, the same rules apply in pattern making and spruing systems as in bronze, although, because of the brittle nature of grey cast iron, thicker castings are recommended and gates can be at the bottom of the runner, since dross and dirt floats on top of iron trapping them in the upper part of the runner. In grey cast iron, carbon (graphite) is suspended as flat, planar crystals throughout the matrix of iron. When stressed, the iron breaks along these planes. In ductile iron, the graphite is grouped in nodules with pure (and ductile) iron in between the nodules. When stressed, ductile iron bends rather than breaks.

Iron is melted in either the cupola or the cupolette. The cupolette is a short, fat, batch melting version of the cupola with a lid. The cupola is charged continuously, while the cupolette is charged intermittently. Both furnaces consist of a round refractory lined shell with three areas: the well, where the molten iron collects as it melts, the windbox (melt-zone) where the air, which burns the foundry coke (coal which has been reduced to about 98% pure carbon), is introduced via the tuyeres,

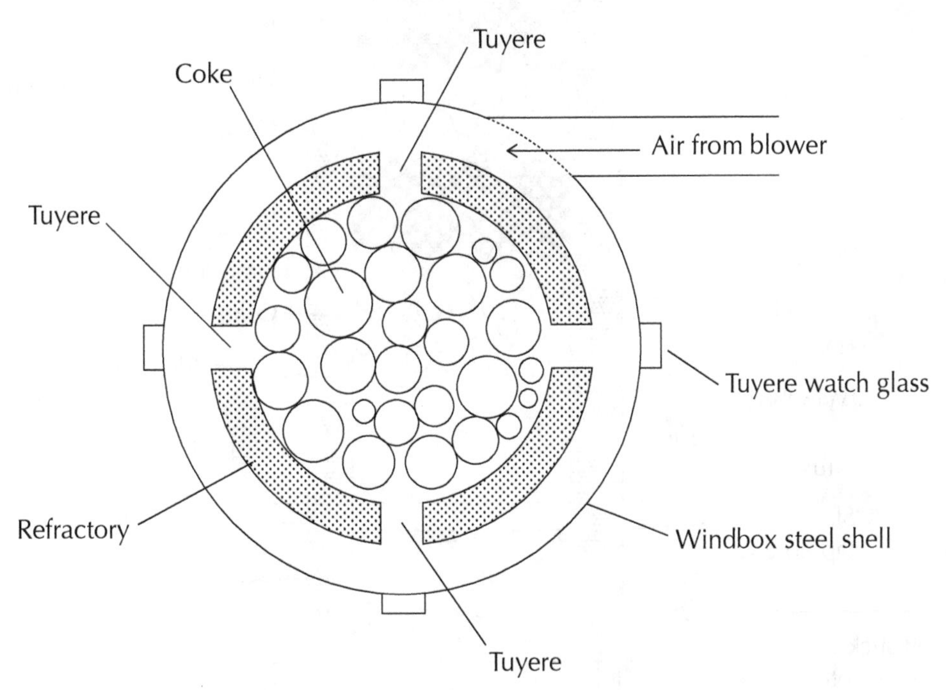

Windbox Cross-Section

and the stack, where the charges of metal are preheated. Coke is piled up inside the furnace to a height above the tuyeres in the windbox. The coke is lit traditionally by building a wood fire in the well and then adding coke on top or with a gas-fired torch until it is burning above the tuyeres. Once the coke is burning above the tuyeres, the blower is then turned on and from then on, the furnace fires on coke and air. More coke is added to establish the "bed height." When this coke is burning the metal charges are added. The metal melts and drips down through the hot coke and collects in the well. The tap hole is plugged with a clay plug called a "bott."

When the well is filled, the bott is picked out, opening the well, and allowing the molten iron to flow out into a ladle.

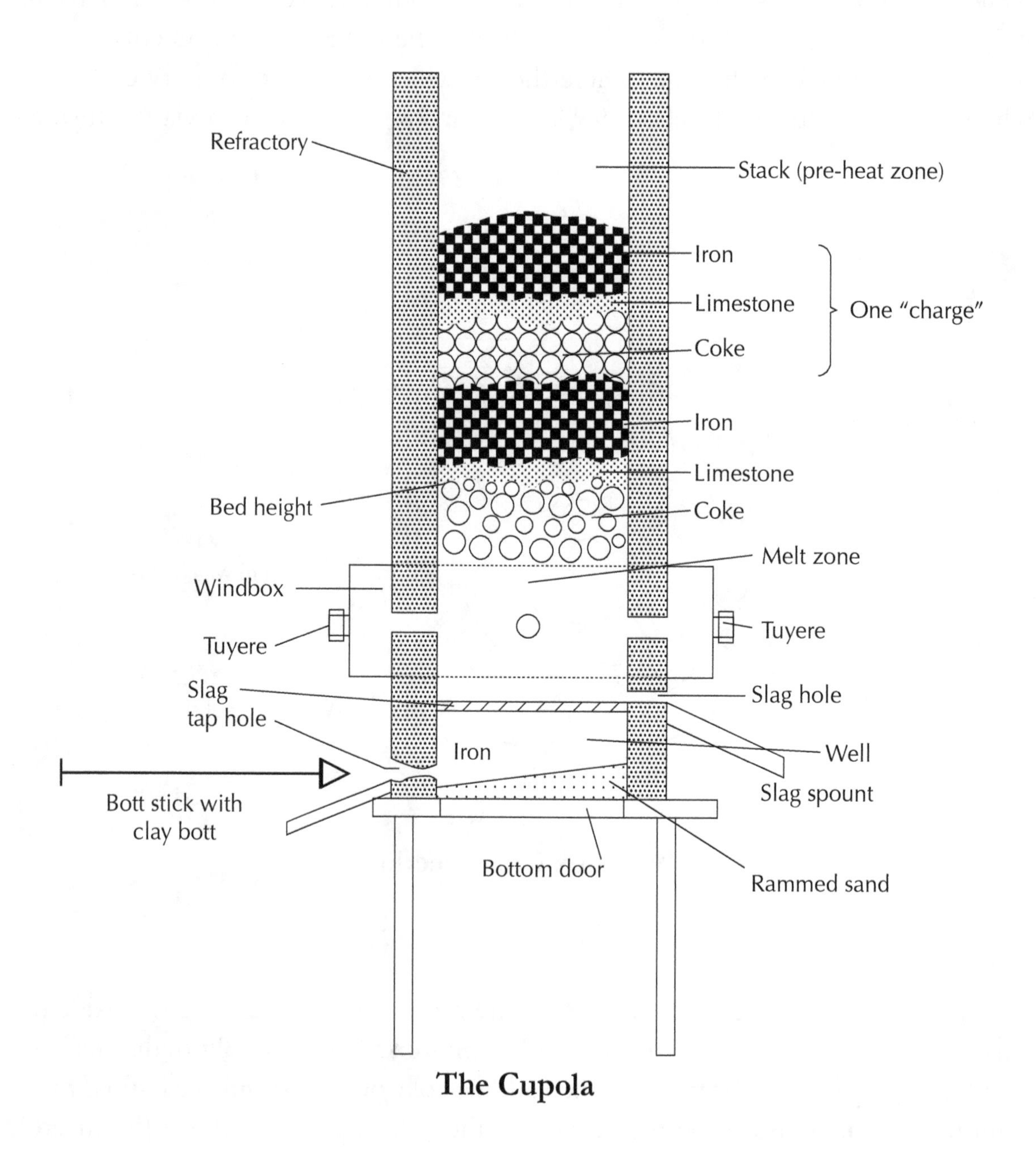

The Cupola

The charges are made up of measured amounts of iron, coke, limestone, and alloying ingredients like copper and ferrosilicon (75 pounds or 41 kgs. of iron, 12 pounds or 5.5 kgs. of low-sulfur coke, 3/4 pounds or .3 kgs. of limestone, 1 pounds or .45 kgs of copper, and .5 pounds or .23 kgs of ferrosilicon for a 12" ID cupola). The charge order is always coke first, then iron, then flux and alloying materials. The size of the iron pieces and the coke are dependent upon the inside diameter of the cupola (approximately 1/10th the ID). As the coke is consumed in the windbox area, it is replaced by more coke falling down from the charges in the stack. As the iron melts, it flows through the melted limestone (slag) which floats on the molten iron and cleans and purifies the iron. When the level of the iron has risen to the level of the slag hole, the slag flows out and is an indication that it is time to tap out the iron from the well. Cupolas can also be designed with 12" in diameter in the stack and windbox areas and 15" in diameter at the well, allowing 300 pounds of iron to accumulate before it becomes necessary to tap out. It takes about 20 minutes for the cupola to melt 300 pounds of iron.

Running the cupola is very labor intensive, as the iron must be broken up into small pieces, the charge materials weighed and organized, the cupola continuously charged and tended, and metal tapped and poured every 20 minutes. Since the mid 1980s, the cupolette has become the more frequently used machine for melting iron by sculptors who do their own foundry work . Because it is a batchmelter, it is easier to manage, can be charged with the amount of metal needed at one time, and requires less people to run.

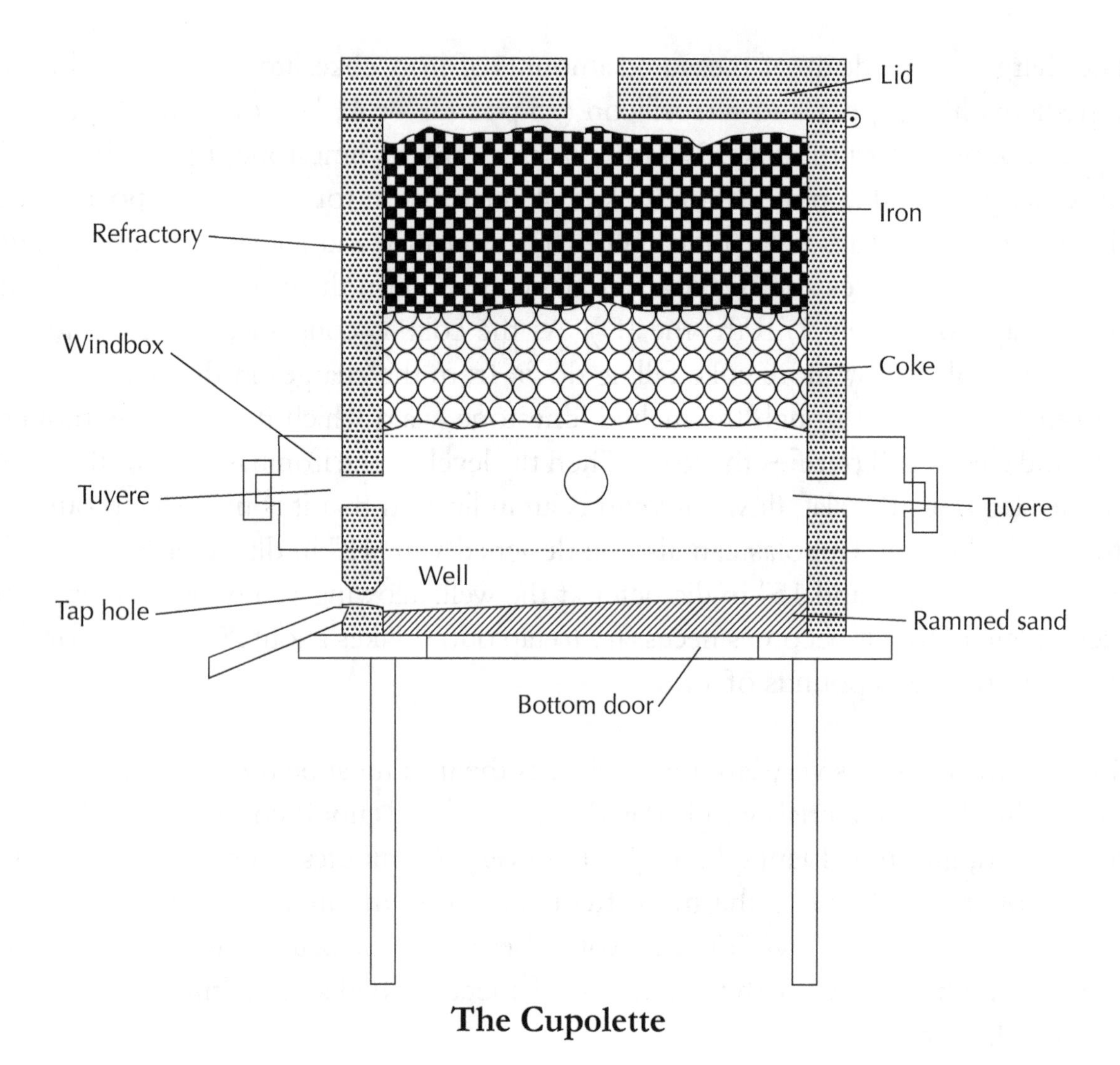

The Cupolette

The cupolette works similarly to the cupola but differs in that it has a lid and is charged with the full amount of metal capacity (or a lesser amount) at one time instead of continuously. The lid acts as the stack of the cupola and keeps the oxygen confined in the melt-zone so the coke will burn. The cupolette at the University of Minnesota Sculpture Foundry has an inside dimension of 24" [61 cms.] in diameter and will melt 300 pounds [136 kgs.] of metal in 20 minutes. The cupolette can be easily run with just four persons. A very popular version of the cupolette has a 12" ID [30.5 cms.] and produces about 100–150 pounds [45–68 kgs.] of iron every 20 minutes.

Prof. Cliff Prokop (Keystone College, La Plume, Pennsylvania) and artist Chris Dashke designed the first cupolette in the early 1980s from an old plan called the Shaw-Walker cupolette. There are lots of other versions of iron furnaces, including continuous tapping cupolas and strange hybrids between cupolas and cupolettes. The Cast Iron Movement, since the first Conference on Contemporary Cast Iron Art (held in Birmingham, Alabama, at Sloss Furnaces National Historic Site in 1988) has spawned numerous other conferences, workshops, and cast iron installations across the country and indeed internationally. See the websites www.internationalfe14.com/and www.wciaa.org/ for more information.

For design specifications for cupolas and cupolettes, see the *Appendix H* and *I*.

Melting Procedures for Iron

Prior to the pour the cupola or cupolette is cleaned and patched. The cupola comes apart in three sections (well, windbox, and stack) for easy cleaning. The cupolette, being shorter, does not have to be taken apart to be cleaned and patched. Sections of both the cupola and cupolette are sealed with a clay gasket (33% fireclay and 67% silica sand). The bottom door is secured and a bed of bottom sand (coarse silica sand 97% and western bentonite 3% by weight is mixed with vermiculite to "taste" with water to make a "squeeze tested" mix) is rammed in so that the bed slopes slightly from the rear of the well to the tap hole. Pay particular attention to ramming around the edges of the well to make strong seal. If lighting with a torch, place a piece of 3/8" [.95 cm.] plywood or pine board on the bed inside the well in front of the tap hole; the board protects the bed during the initial lighting of the bed coke.

Large pieces of coke about the size of large softballs are placed inside the well to just below the tuyeres. The area just inside the tap hole should be a tunnel of coke so that the flame will penetrate into the middle of the coke bed. Place a "stopper" piece of coke at the end of the tunnel in the middle of the well; this will help spread the flame up and around the coke bed. Most of the time the front of the coke bed (toward the tap-hole) lights last. Continue adding the smaller coke until the bed height is established. (Consult *Appendix J* for bed height calculation.) Building a wood fire inside the furnace and then adding the coke is the traditional way to light the coke.

Spread damp dirt around the base of the cupola or cupolette stand. The gas torch is positioned at the tap hole and ignited so that the flame burns through the tap hole and into the well. When the entire bed of coke is burning (bright yellow-orange color at the tuyeres), the torch is removed and the air turned on. Run under blast for 15–30 minutes and then adjust bed height before charging any iron. Accurate notes should be kept to compare the action of the melt with previous ones. It is better to add more coke and thoroughly burn in the bed rather than add iron too soon. See *Appendix K* for a typical iron pour record sheet.

The cupolette has a lid that must be opened to charge, so be sure that the cupolette is in a well-ventilated area. Cupolas inside are usually stacked right out through the roof. The metal is charged by the next crew immediately after the tap hole has been botted following a tap-out. The charges for the cupolette are larger than for the cupola and can be adjusted to the molds that are being poured. For a 24" ID cupolette, usually the amount charged is 250–300 pounds [113–136 kgs] of iron. Coke charges are not weighed, but coke is added to a gauge height that corresponds to about 4" of coke above the calculated bed height. Four inches of charge coke added above the bed height is usually sufficient to melt any iron charge in any sized cupolette. Using a gauge that fits over the edge and into the inside of the cupolette helps facilitate accurate coke charging without weighing each charge. Rake the coke charge flat and then add the measured amount of iron, copper, ferrosilicon, and limestone.

When the first charge is added, the time is noted. Iron should be dropping by the tuyeres steadily in about 4–5 minutes. Iron sooner than 4–5 minutes means the bed height is too low; iron later than 5–6 minutes means the bed is too high. Adjust the bed height and bott up the cupola or cupolette when a thin steady stream of iron runs out of the tap hole. The bott mix is 33% wood flour (or screened sawdust), 33% silica sand, and 34% fireclay with enough water to form a moldable consistency.

During the first few heats, you can run a little oxygen (about 1/2 or 1 pound of pressure) into the windbox to boost the heat inside the melt-zone. Usually the first several taps of the furnace result in colder metal. As the cupola or cupolette heats up, the temperature of the iron will also get hotter at the tap. A male air coupling is welded into the windbox; a hose with a female air coupling on one end and runs from an oxygen bottle (stored securely away from the furnace) to the male coupling on the windbox. The quick disconnect allows removal of the hose before the tap-out. About 1/2 pound of oxygen is trickled in during the heat. Remember to turn off the oxygen while tapping and charging. Running small amounts of aluminum in the charge metal will also boost the temperature during the first heats.

In about 15–20 minutes the well should be full of iron and ready for tap-out. Usually the slag hole is opened at the 12–15 minute mark to check to see if slag is flowing at the slag1 hole level. The slag hole is botted up when iron begins to flow out of it. Iron at the slag hole is a signal that the well is nearly full and should be tapped soon. Slag that is thin, very viscous, and flows easily is a good sign.

More than 100 pounds of iron should be tapped out into a large ladle supported by a bail and a bridge crane. With the ladle in position, a pick is used to remove the clay bott and the iron runs into the ladle. The temperature is read and the big ladle is skimmed of slag. Use of a "slag grip" or "popcorn" is recommended to consolidate slag. If there are many small molds, a portion of the iron is transferred to hand ladles and poured by a four-person crew. Large molds are poured directly from the big ladle in the area covered by the crane. Both ladles should be kept hot between taps. Exothermic ladle liners are a good investment—no need to heat the ladle, as when refractory is used to line the ladle. They come in various standard sizes to fit various ladle sizes; various manufacturers make a variety of models including self-skimming ladles. Care must be taken when skimming with a metal skimmer so as not to damage the fibrous wall of these liners and care should be taken to drain excess metal completely to avoid sculling.

When the ladle is full, bott the tap hole before any excess slag (that is not run off at the slag hole) runs into the ladle. The use of bott sticks is recommended. A bott stick with a small rim to hold the clay bott and a small short spike in the center is a good design. Botts should be rolled in sea coal to facilitate removal. The problem with hand botting (slapping the bott onto the tap hole), is that one, you never see this practice in industry, and two, the cone shape of the bott on the bott stick fits the cone shape of the tap hole making a good seal and eliminating excess bott material. Tap holes should be designed so that they are like two cones, one outside and one inside, separated by about 1" [2.5 cm.] length of the hole. A tap hole of 1" in diameter tapering to 2–3" [5–7.7 cm.] on the outside and inside works well and stays hot on the inside.

It is recommended that the air blast be turned off when tapping out and charging the cupolette; cupolas are charged continuously (as each charge is consumed and the level at the top of the stack drops), but it is recommended that charging stop during the tap-out. After all the iron is poured, the air blast is turned off and the bottom door is dropped. If all goes well, the sand bottom, the coke, and a small amount of iron will also fall out. The bottom sometimes must be coaxed with a pick-axe or a levered pick called the "Linden Pick-o-matic." Hose down the dropped bed of coke and iron with water so that it does not harm the legs of the well platform. Be careful of the steam that will be generated with this procedure.

Because of the higher temperature of iron, alkyd oil and furan sand molds tend to emit more smoke (especially melt-out molds) and should be removed from an inside pouring area as soon as they have chilled.

Iron that is removed from the mold too early will be harder (and consequently more difficult to chase and finish) than iron left to cool more slowly in the mold. Great care should be taken when shaking out iron castings, as they are significantly easier to break when removing the sand mold. In most jurisdictions, both the sand mold material and shell mold material is classified as hazardous and must be properly disposed. The sand from used sodium silicate molds can be reclaimed by soaking the molds in water to break the silicate bond; other binder sands can be reclaimed by heating to break the bonds, but this is most impractical in small operations.

Induction Melting of Iron

The Inductotherm Tilt Furnace can be paired with the Inductotherm Lift-Swing Furnace and operated from the same control panel. You should be checked out on the safe operation of any induction melting system. Ferrous metals are the only metals melted in the tilt furnace. Thick ingots or chunks of iron may be melted; there is no need to break the iron up as when melting in the cupolette. Ingots and scrap charged after there is a melt in the furnace must be heated up on the gas crucible furnace before being added to the melt. Failure to observe this rule will result in exploding iron as the water in the ingots or scrap steams out of the metal.

Some operations use a crane to tilt the furnace over to pour the iron into a waiting ladle and some tilt furnaces have hydraulic tilting mechanisms. If using a crane to tilt the furnace, it takes four people to operate: pour person, dead-end person, skimmer, and crane operator. The ladle is positioned over the safety pit in front of the tilt furnace. The crane hook is hooked to the rear of the tilt and the crane is raised, tilting the furnace, and pouring the iron into the ladle. Do not raise the crane too high or the furnace will be ripped from its foundations! Always preheat the ladle before adding the molten iron. Wear earplugs while this furnace is in operation. As with the operation of the lift-swing furnace, persons with electronic medical devices cannot be in the room during the operation of the tilt furnace.

Gold is for the Mistress,
Silver for the Maid,
Copper for the Craftsman,
Cunning at his trade.

"Good!' said the Baron,
Sitting in his hall,
"But Iron—cold Iron
Is the master of them all."

Rudyard Kipling

[illegible]

[illegible]

[illegible]

[illegible]

Chapter 9

CHASING

"Chasing" refers to the removal of sprues, vents, and any casting defects and the restoration of form and surface texture to the castings. It also can refer to the addition of detail and texture not found in the original pattern for the casting.

There are three main phases to chasing: rough chasing, where the sprues, vents, and any extraneous metal forms or defects are removed; an intermediate chasing, where the metal is worked so that the form in the sprue and vent areas is contiguous with the rest of the casting; and finish chasing, where scars are blended and texture is restored to those areas where extraneous metal has been removed.

To rough chase, use a bolt cutter, hacksaw, or cut-off disk on a handheld grinder to remove the sprue, vents, risers, or any flashing. Large sprues and gates can also be removed with an Air-Arc with carbon rods hooked up to an arc welder. Place the piece in the vise, making sure that the vise jaws are shielded with a sheet of lead to avoid putting vise-jaw marks into the casting. (Commercially made aluminum vise jaw shields, wood, or several layers of cardboard may also be used.)

Cut or grind slightly above the surface of the form. If you cut too deeply, a much larger area will have to be finish chased. Use a cold chisel to cut away "pearls" or any other small projections. To keep the cold chisel from cutting into the surface of the metal sharpen the chisel according to the following diagram:

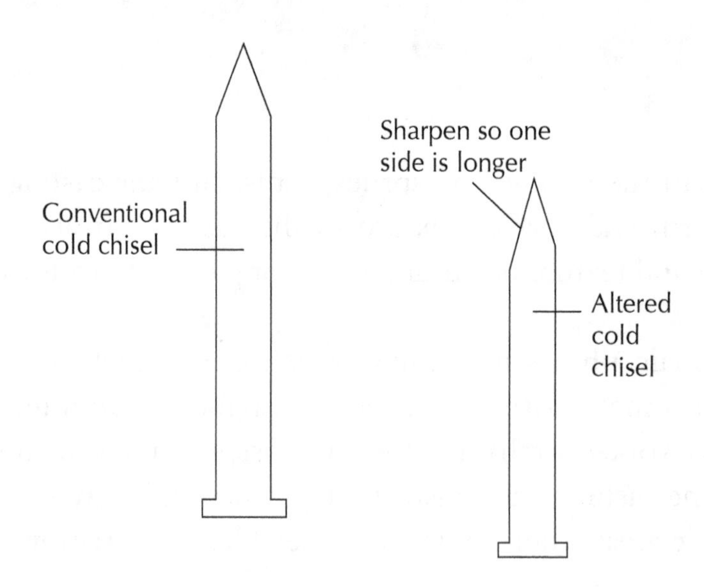

Use the long side against the form of the casting. This sharpening configuration will help keep the chisel from digging into the metal, resulting in an even larger area to chase.

After all the sprues, vents, and other extraneous metal have been removed, return this metal to the proper scrap bin and weigh the rough chased casting. You are now ready to chase the piece so that the form is contiguous and there are no raised scars on the casting.

This intermediate chasing step may be accomplished with rotary bits or flap wheels used in an air or electric die grinder, files (a half-round bastard file is recommended; other file cross-sectional shapes are triangular, round, and square), cold chisel, or handheld mini-grinder with 5" grinding disk or 50 grit sanding disk. The important thing during this stage is to reduce the raised metal area so that the form is contiguous without enlarging the scar too much.

During the finish chasing stage, the idea is to recreate the texture of the surrounding area on the scarred surface and to blur the distinction between the scarred area and the rest of the cast texture. To do this you need to bring something of the castings texture into the scarred area, while also extending the texture you are making into the unscarred area so the eye does not detect the boundary. By hammering points, nail sets, lozenge chisels, cold chisels, old files, rough pieces of iron sprues, old 30–50 grit sanding disks, or other specially created chasing tools against the surface, one can recreate a texture or extend the textural aspect of the piece with new or additional surface. Running a rotary bit at slow speed so that it "chatters" against the surface or using various grit flap wheels can create other textures. An air-powered "needle scaler" is another nice tool for texturing the surface.

The metal surface may be smoothed with fine sanding disks on mini-grinders or on small mandrels that fit into die grinders. Of course, hand sanding with emery paper or crocus cloth will also smooth the top surfaces of a texture. To achieve polished surfaces, one needs to go through the proper sequence of finer and finer grit emery paper and then move to a cloth buffing wheel, beginning first with black emery compound and then continuing to buff with rouge compound. Because the compounds are greasy, it is important to thoroughly clean the buffed surface before attempting to patina.

After the piece is finished chased, make any hanging devices, tapped holes, or other alterations to the form before sandblasting in preparation for painting, patinization, or other surface treatment. If your casting must fit to some other form, make sure that everything works before doing the patina.

Proper chasing is extremely important to the final look of the metal casting. Badly chased surfaces will stand out when patinated. It is important to allow adequate time and patience for chasing as it often takes as much time as the mold making and casting. As the old foundry curmudgeon has often remarked:

"After making the pattern, completing the mold, melting the metal, pouring, and shaking out the casting, you are half done!"

Patina of Metals

Chapter 10

Patination of metals involves the chemical coloring of the surface. Aluminum is the most difficult metal to patina because it forms its own tough oxide that resists chemical action. Iron colors can range from black through blue and green to brilliant oranges and reds. Bronze is the metal most interesting to patina because of the wide variety of colors and effects possible.

Before beginning any patina process, make sure that the casting is properly prepared. The piece must be absolutely finish chased. Since patina is the last step, any mechanical changes to the surface of the casting will alter the patina. Patinas do not hide bad finishing or chasing, if anything they amplify any casting defects. All mechanical systems for presenting the piece, such as mounting systems, holes tapped for bolts, hinging systems, etc. must be functioning. If you are mechanically joining a casting to another casting or other material, try it out first prior to the patina application. The piece must be clean of oil, grease, or paint. Since most patinas are water-based, any oil or grease will resist the patina. If the piece has oil or grease, clean it with a de-greaser such as a strong soap, then rinse it clean with water. Lightly sandblast the piece to give it an even all-over texture. Sandblasting will not destroy your texture unless you overdo it or the texture is extremely subtle. Avoid touching the casting with your fingers after sandblasting; from this point on use disposable rubber gloves when handling the casting.

There are many variables that will affect how a patina will look on any casting. The composition of the metal alloy will affect the color, as will the purity of the chemicals and water used. The method of application and the sequence of methods is very important, especially for the richer looking patinas. The recipe or formula used and measurement of the ingredients is the main factor in determining whether the color described will come up on your casting. Of course experience and patience are two virtues needed by everyone attempting patinas—Many patinas develop slowly and the rush to get a fast color may result in a bad one. Experimenting on test pieces is one way to gain experience and also satisfy the urge to experiment without jeopardizing a finished piece. Remember: "Better" is sometimes the enemy of "Good."

When mixing chemicals, wear eye and hand protection. Avoid wasting or spilling the chemicals. Clean up and wash down all spills or leaks! Chemicals should be stored in a dry, well ventilated and secure area. Be sure to seal each jar tightly and turn to the appropriate shelf. Be sure to label all mixed solutions and spray bottles. Labeling of all chemical containers is an OSHA requirement. DO NOT fail to label what you mix; all of the ingredients must be on the label. All patination should occur in a well-ventilated patina area. If using an acetylene/oxygen heating rosebud heating tip, be sure it is clean and not corroded with chemicals. Clean the torch orifices with a tip cleaner if they are clogged or if you do not get a steady flame during light-up. Be sure your sink or water bath is clean. Fill the sink with clean, cold water for use as the rinse for the pieces during the patina process. Wear your mask rated for mists, safety glasses, and disposable latex gloves during all patina operations.

Methods of Patina Application

1. **Immersion-** Place the piece in a hot or cold patina solution so that the solution completely covers it. This method works best for small parts or pieces. This method is impractical for larger works because of the need for large glass or plastic containers and the need to mix large amounts of patina chemicals. Most immersion methods works in 24 hours or less.

2. **Atmospheric-** Simply placing the piece in a natural (or industrial) atmosphere will color bronze or iron. The green color seen on copper-clad church roofs is the result of chemicals in the air. Of course, this method is not practical for those trying to meet a deadline. One can construct a closed atmosphere with lidded plastic buckets or make enclosures for larger pieces with plastic sheeting and duct tape. Place a fuming solution (such as ammonia) inside the bucket or tent and seal. Do not allow the solution to come in contact with the piece. Clear plastic allows you to view the action of the fumes on the surface of the metal. Always wear a respirator rated for fumes when working with atmospheric

patinas. Most atmospheric patinas (except those in the great outdoors) will work in a few days.

3. **Burial-** The piece is buried in sand, dirt, sawdust, or manure which has been prepared with patina solutions and left for days, weeks, months, years, or centuries. Long burial patinas are impractical for the artist attempting to meet an exhibition deadline. Wrapping the piece with patina-solution-soaked cloth, straw, rope, leaves, grasses, or other porous material not only produces colors, but also makes for interesting textures in the patina. Wrapping patinas can usually work as early as one or two days.

4. **Spray-** The piece is heated to 200° F [93° C] and then the solution is sprayed on with plastic spray bottle using the finest setting. This method is the most economical, as not much solution is wasted. Be sure to wear your mask rated for mists and use disposable rubber gloves when handling the piece. Continue heating the piece until the moisture evaporates, but does not boil off of the surface of the piece and then spray again. Continue spraying and heating until the color desired is achieved. Rinse between repeated applications or between applications of different solutions.

5. **Brush-** Use a natural bristle brush, sponge, roller, or rag to brush, daub, or wipe the patina solution on to the piece. The piece should be heated to 200° F [93° C] between applications and the solution should evaporate off of the surface. The brush method is especially useful in applying patina to selected areas on a piece. Rinse between repeated applications or the application of another solution.

6. **Heat-** Heat the piece with a torch, furnace, or fire until the desired color has been achieved. Heat the piece to 500° F [260° C] and cover it with organic material such as sisal, leaves, grasses, or straw. Smoke the piece by suspending it above burning sisal, leaves, or wood. (This method can produce a very rich black.)

7. **Electric-** The piece is electroplated with nickel, gold, silver, or copper; the piece is anodized if it is aluminum. These processes require very good castings and are usually done in commercial establishments. Natural electroplating or "strike" plating aluminum or iron with copper is easy to do with a saturated solution of copper sulfate, water, and weak sulfuric acid.

"Patina" can also mean the application of oil paint, enamels, oil stick, pastels, powdered pigment, or oil stains to the surface of the metal. Paint can be combined with traditional chemical patinas. Colored waxes can be used to change or enhance the color of a patina. There are many commercial solutions such as rust-proofers or gun bluing which will patina metal. A rich black can be obtained by heating the metal to a high heat and brushing motor oil, linseed oil, olive oil, molasses, or sugar solution till the color you want appears. The patina you are looking for is the one that enhances both the form and the concept of the piece.

With most water-based patinas, rinse the piece in water following the final application of color. Allow the piece to dry thoroughly. A patina left unsealed will continue to "work" in the atmosphere and during changes in humidity. Wax is the traditional sealer for indoor pieces. Heat the piece in an electric oven to 250° F [121° C] until all the water has evaporated from the piece. A wax with a high carnauba content is a good, tough wax, but tends to change patinas. The patina will darken with the application of these waxes and will look, when waxed, much like it did when rinsed and wet. "Kiwi" brand neutral shoe wax is a good choice because it will not change the color of the patina as radically as other waxes. "Renaissance Wax," available from some hardware outlets or online, is the superior wax for color retention. Melt the wax carefully in a clean container and brush the hot wax onto the hot patinated piece. Do not apply the wax too heavily. Allow the wax to cool and then rub gently with a clean soft cotton cloth. Steel wool may be used to bring up highlights or reveal the true color of the metal.

Outdoor pieces can also be treated with wax or, for a more permanent treatment, with lacquer or plastic finishes. Re-waxing or refinishing of outdoor pieces is required every 2–4 years; consult the Bibliography for additional information.

There are literally thousands of patina recipes for bronze, less for aluminum or iron. See *Appendix F* for traditional recipes or consult the Bibliography for additional texts on patinas.

Enameling of Metals

Both bronze and iron can be enameled to achieve rich, bright, and glassy colored surfaces. Enamels are essentially powdered glass-like materials that are fused to metal with heat. They are formulated to work with particular metals, so be sure to consult the supplier for the appropriate enamel. For bronze, low temperature, high expansion enamels are used. The surface of the bronze or iron must be clean, dry, and sandblasted. A solution of gum arabic is painted on the surface to be enameled and then the enamel is applied with a shaker equipped with a fine sieve. The gum arabic is allowed to dry and the piece is then placed in an oven until the enamel melts and fuses to the metal surface. The temperature for most enamels for bronze is 1050–1100° F [565–593° C]. Enamels can also be applied directly to hot metal surfaces (which is the way most sinks and tubs are enameled), but this is a more difficult method without special handling equipment and kilns.

A wide variety of enamels for bronze, copper, and iron and excellent technical assistance is available from the Thompson Enamel Co.
[http://www.thompsonenamel.com/]

Electroplating

The color (and to some extent, the texture) of a cast surface can also be changed by plating the surface with copper, tin, chromium, nickel, cadmium, or gold. Most electroplating is done by commercial companies, however both copper and nickel can be plated in the studio with easily built equipment, provided proper safety procedures are followed. Electroplating takes place in a bath of salts compatible with the metal to be plated. The object to be plated is the cathode (negative pole) of the electric current and a bar of the plating metal is the anode (positive pole) of the system. When an electric current (DC) is passed through the solution, the plating metal passes from the anode through the solution and is deposited in an even layer on the cathode (object). Most plating is very thin and follows the surface texture of the object, but by varying the amperage of the current, a fast, heavy plate will build up, creating a texture. A convincing light-weight "bronze" can be produced by plating an aluminum casting with copper.

A simple device to plate copper can be made from a battery charger or DC transformer, a rheostat, ammeter, and voltmeter. The plating is done in a plastic bucket. Be sure that the plating area is adequately ventilated, as this process can produce harmful fumes. The electrolyte solution should be about 70° F [21° C]. Use a low amperage to produce a fine-grained plate; a high amperage will produce a rougher, more porous, and spongy plate whereas too high of an amperage setting will cause a darkened and burnt deposit.

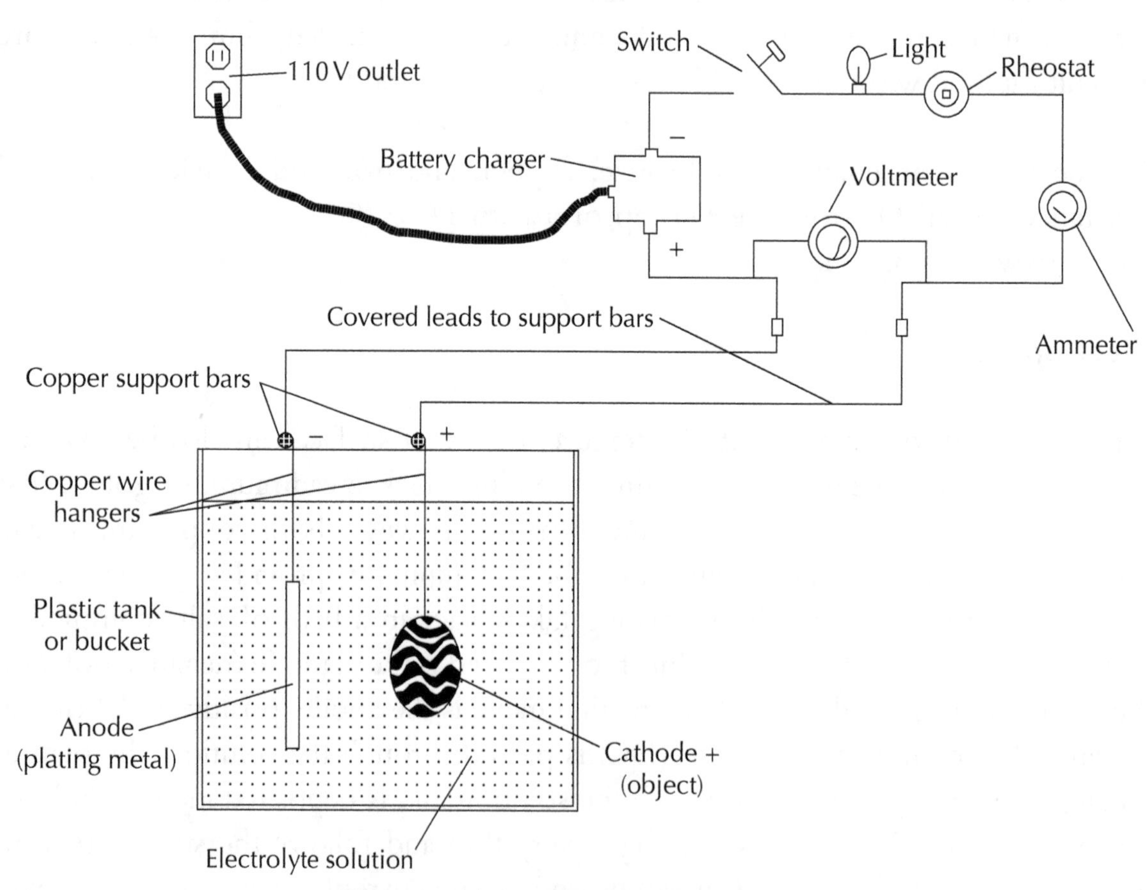

Electroplater Wiring Diagram

For copper plating, to make 1 gallon (about 4 liters) of electrolyte solution:

1. Dissolve 1.75 pounds [.8 kgs.] of copper sulphate in 1 gallon [4 liters] of distilled water, then
2. Add 3.5 ounces [.01 kgs.] of sulfuric acid.

Less acid and low current produces a soft plate, while more acid and a higher current will result in harder deposits. A rosy color on the metal indicates a satisfactory plating action.

For nickel plating (fast plate):

1. Dissolve 2 pounds [.9 kgs.] of nickel sulphate,
2. 4 ounces [.12 kgs] of ammonium chloride, and
3. 4 ounces [.12 kgs] of boric acid in 1 gallon [4 liters] of distilled water.

A low amperage will result in a hard brittle deposit, while a current that is too strong will produce a dull plate which can separate from the base metal. A bluish color indicates a good plate.

Iron cannot be plated directly with copper as the sulfuric acid reacts with the iron causing the copper to lift off of the iron. Iron should be nickel plated first and then plated with copper. Iron can be "strike plated" with copper by dipping it for a short time in a saturated solution of copper sulfate and sulfuric acid.

Nonmetallic objects can be plated if they are first covered with a conductive paint or other conductive surface such as metal foil. Silver paint such as Dupont 4817 is an excellent conductive since silver will conduct electricity in the electrolyte solution even though tarnished. Other paints such as copper print paint will work, but must be bright.

Zinc plating is generally done commercially by hot-dipping the object in a molten bath of zinc. Small pieces can be plated by dipping them in a crucible of molten zinc. Care must be taken not to overheat the zinc to prevent it from catching fire.

Painting

Both iron and aluminum are good metals to paint to achieve a final finish, though there are no rules against using paint on bronze. The metal surface must be dry, and free of dirt and grease. Sandblasting is recommended to provide a tooth for the paint. Use oil-based enamels, artist's oil paint, or model paints; latex and acrylic paint are not recommended. Primers containing zinc are good for a base coat on iron. Of course, how the layers of paint are applied is a question of how the paint will enhance the form and clarify the content of the piece. Nice effects are achieved by leaving initial layers of pigment showing through subsequent layers. Oil stains and oil-based pastels are another way to color the surface of the casting. For those interested in the art of painting, painting on a three-dimensional surface presents a significant additional challenge.

Remember that "finished" means the most appropriate surface, texture, and color to reveal both your formal statement and the conceptual content of the work. It does not mean that the piece needs to look like Danish furniture or a Southern California hotrod and it does not mean that the finish interferes with the mind's and eye's reading of the piece. Rather, it is a harmonious conclusion to the artist's process and statement.

Bibliography

History

Agricola, Georgius. *De Re Metallica.* (Translated by Herbert Hoover) First printed in 1556. Dover, 1950.

Barnard, Noel. *Bronze Casting and Bronze Alloys in Ancient China.* Australian National University, 1961.

Biringuccio, Vannoccio. *Pirotechnia*, First printed in 1540. M.I.T. Press, 1966.

Cast Iron and the Crescent City. Exhibition Catalog. Gallier House, 1975.

Hauser, Caroline. *Greek Monumental Bronze Sculpture of the 4th and 5th Centuries B.C.* Garland, 1987.

Mattusch, Carol. *Casting Techniques of Greek Bronze Sculpture.* University of North Carolina, 1975.

National Park Service. *Hopewell Furnace.* National Park Handbook 124. National Park Service, 1983.

Neaher, Nancy. *Bronzes of Southern Nigeria.* Stanford University, 1976.

Raymond, Robert. *Out of the Fiery Furnace.* Pennsylvania State University Press, 1986.

Savage, George. *A Concise History of Bronzes.* Praeger, 1969.

Shapiro, Michael. *Bronze Casting and American Sculpture.* University of Delaware, 1985.

Simpson, Bruce. *History of the Metalcasting Industry*, 2nd edition. American Foundryman's Society, 1969.

William, Watson. *Ancient Chinese Bronzes.* Faber and Faber, 1962.

Safety

Rossol, Monona. *The Artists Complete Health and Safety Guide.* Allworth Press, 2001.

U.S. Department of Health, Education, and Welfare. National Institute of Occupational Safety and Health [NIOSH]. *Health Hazards in a Foundry.* NIOSH Publication 77–103, "Melting and Pouring." DHEW, 1976.

———. *Health Hazards in a Foundry.* NIOSH Publication 77–102, "Pattern Shop, Core Room, Molding Shop, and Sand Handling." DHEW, 1976.

———. *Health Hazards in a Foundry.* NIOSH Publication 77–104, "Shakeout, Cleaning, and Grinding." DHEW, 1976.

———. *An Evaluation of Occupational Health Hazard Control Techniques for the Foundry Industry.* NIOSH Publication 79–114. DHEW, 1978.

General Sculpture Technique

Andrews, Oliver. *Living Materials.* University of California, 1983.

Clark, Carl. *Modeling and Casting.* 2nd Edition. Standard Arts Press, 1946.

Rich, Jack C. *The Materials and Methods of Sculpture.* Oxford, 1947.

Untracht, Oppi. *Metal Techniques for Craftsman.* Doubleday, 1968.

Verhelst, Wilbert. *Sculpture: Tools, Materials, and Techniques.* 2nd Edition. Prentice Hall, 1988.

General Foundry Technique

American Foundry Society. *Copper-Base Alloys Foundry Practice.* AFS, 1965.

American Foundry Society. *Foundry Sand Handbook.* 7th edition. AFS, 1963.

American Foundry Society. *No Bake Cores and Molds.* AFS, 1980.

American Foundry Society. *Patternmaker's Manual.* 2nd edition. AFS, 1986.

Ammen, C. W. *Casting Brass.* Tab Books, 1985.

Ammen, C. W. *Casting Iron.* Tab Books, 1984.

Ammen, C.W. *The Metalcaster's Bible.* Tab Books. 1980.

Brown, John. *Foseco Foundryman's Handbook.* 11th edition. Butterworth-Heinemann, 2000.

Clark, Carl. *Metal Casting of Sculpture.* Standard Arts Press, 1948.

Cowles, Fred. *An Elementry Foundry Manual.* McEnglevan Heat Treating and Manufactoring. 1967.

Hamilton, Ed. *Patternmaker's Guide.* AFS, 1976.

Hitchcock, Howard. *Out of the Fiery Furnace: Casting Sculpture from Ceramic Shell Molds.* Kaufman, 1986.

Jackson, Harry. *Lost Wax Bronze Casting: A Photographic Essay on This Antique and Venerable Art.* Van Nostrand, 1979.

Mills, John. *The Technique of Casting for Sculpture.* Batsford, 1990.

Montagna, Dennis. *Conserving Outdoor Bronze Sculpture.* National Park Service, 1989.

Sanders, Clyde A. *Foundry Sand Practice.* American Colloid Company, 1973.

Young, Ron D. and Robert A. Fennel. *Methods for Modern Sculptors.* Sculpt Nouveau, 1980.

Patination

Edge, Michael. *The Art of Patinas for Bronze*. Artesia Press, 1991.

Fishlock, David. *Metal Colouring*. Draper, 1962.

Hughes, Richard and Michael Rowe. *The Coloring, Bronzing, and Patination of Metals*. Van Nostrand Reinhold, 1983.

Young, Ron D. *Contemporary Patination*. 4th edition Sculpt Nouveau, 1997.

Appendix A: Molding with Silicon Caulking Compound

Making a Mold with Silicon Caulking Compound

Tube silicon caulking compound is a cheap alternative to more expensive silicon rubbers for casting wax patterns. Silicon rubber molds are a good choice to mold rigid patterns that do not have draft. Be sure that the tube says that it is 100% silicone, as other mixtures may not work as well. Remember that this silicon is adhesive, so the use of a good release is most important. Seal porous patterns (such as wood, plaster, or fired ceramic) with lacquer or shellac and then use petroleum jelly for the release agent. Use a soft brush to coat the pattern surface with the petroleum jelly. Do not apply the petroleum jelly too heavily or you will change the texture of the pattern. Wet clay does not require a release (but the silicon caulk will need a longer drying time) however, use a release with Plasticene.

Place the pattern on a flat nonporous surface with plenty of room to accommodate the silicon mold and the plaster mother mold to be made over the silicon mold. The silicon caulk can be thinned with Xylol (also called Xylene). Xylol is a strong solvent; use eye protection, mask, and gloves. The silicon caulk itself has a strong odor due to the release of acetic acid during curing. Use both materials in a well-ventilated area. Avoid contact with skin as it may irritate the skin, nostrils, or eyes. Thin the silicon with about 10% Xylol by volume of the silicon. Mix in paper containers as both materials will eat plastic and Styrofoam. Apply a thin coat of the thinned silicon caulk to about a thickness of 1/8" [.32 cms.] using a disposable natural bristle brush. Be sure the silicon extends beyond the pattern edge by about 1" [2.54 cm.]. Place in a well-ventilated area and allow to dry for 24 hours. Apply a second thin coat of the caulk directly from the tube (or thinned if desired) and then lay a single thickness of cheesecloth or gauze on the mold and work the silicon into the cloth. Allow the silicon to dry for 24 hours. Make a plaster mother mold, coating the cured silicon mold to a thickness of 1" [2.54 cms.]. Depending on the form of the pattern, this may need to be a two-(or more) piece mold. After the plaster has cured (it will be hot to the touch), remove the mother mold and gently pull the silicon mold from the pattern. Store the silicon caulk mold in the mother mold when not in use.

Appendix B: Weight of Pattern to Weight of Casting

A pattern weight of 1 lb. or .45 kgs made of	Will weigh in **Iron**	Will weigh in **Bronze**	Will weigh in **Aluminum**
Birch	10.6 lb. / 4.8 kg	12.3 lb. / 5.6 kg	3.9 lb. / 1.8 kg
Oak	9.4 lb. / 4.2 kg	10.8 lb. / 4.9 kg	3.4 lb. / 1.5 kg
Pine	14.4 lb. / 6.5 kg	16.6 lb. / 7.5 kg	5.3 lb. / 2.4 kg
Cardboard	35 lb. / 15.8 kg	39 lb. / 17.6 kg	13 lb. / 5.9 kg
Pressboard	7.5 lb. / 3.4 kg	8.0 lb. / 3.6 kg	2.5 lb. / 1.1 kg
Plasticene	5.0 lb. / 2.3 kg	5.5 lb. / 2.5 kg	1.85 lb. / .84 kg
Sand	5.0 lb. / 2.3 kg	5.3 lb. / 2.4 kg	1.6 lb. / .73 kg
Cedar	12.5 lb. / 5.7 kg	14.7 lb. / 6.7 kg	4.5 lb. / 2.0 kg
Mahogany	8.5 lb. / 3.8 kg	10.0 lb. / 4.5 kg	3.1 lb. / 1.4 kg
Wax	8.5 lb. / 3.8 kg	10.0 lb. / 4.5 kg	3.1 lb. / 1.4 kg
Lead	.64 lb. / .29 kg	.74 lb. / .33 kg	.23 lb. / .10 kg
Zinc	1.0 lb. / 4.5 kg	1.17 lb. / .53 kg	.36 lb. / .16 kg
Masonite	6.7 lb. / 3.0 kg	8.0 lb. / 3.6 kg	2.6 lb. / 1.2 kg
Plaster	4.0 lb. / 1.8 kg	4.7 lb. / 2.1 kg	1.6 lb. / .73 kg
1 oz of Pink Styrofoam or .028 kgs	25 lb. / 11.3 kg	29 lb. / 13.1 kg	9.9 lb. / 4.5 kg

Bronze= ....309 lb./cubic inch or .14 kgs/ cubic cm.
Iron= ....260 lb./ cubic inch or .12 kgs/ cubic cm.
Aluminum=........... .093 lb./ cubic inch or .04 kgs/ cubic cm.
Copper= ....317 lb./ cubic inch or .14 kgs/ cubic cm.
Tin= ....262 lb./ cubic inch .11 kgs/ cubic cm.
Silver= ....5.53 Troy Ounces/ cubic inch or .17 kgs/ cubic cm.

Shrinkage of castings per linear foot or 30.48 centimeters

Aluminum..............3/16" or .48 cm.
Iron........................1/8" or .32 cm.
Bronze...................5/32" or .39 cm.
Tin..........................1/2" or 1.27 cm.
Zinc........................5/16" or .8 cm.
Lead.......................5/16" or .8 cm.

Appendix C: Alkyd Oil Resin Binder and Catalyst Percentages

# of Sand	1% Resin	Catalyst 20% of Resin	1.5% Resin	Catalyst 20% of Resin	2% Resin Catalyst	Catalyst 20% of Resin
50	.5# 8.0 oz	.10# 1.6 oz	.75# 12.0 oz	.15# 2.4 oz	1.0# 16.0 oz	.2# 3.2 oz
100	1.0# 16.0 oz	.20 # 3.2 oz	1.5# 24 oz	.30# 4.8 oz	2.0 32.0 oz	.40# 6.4 oz
150	1.5# 24.0 oz	.30# 4.8 oz	2.25# 36.0 oz	.45# 7.2 oz	3.0# 48.0 oz	.60# 9.6 oz
200	2.0# 32.0 oz	.40 # 6.4 oz	3.0# 48.0 #	.60# 9.6 oz	4.0# 64.0#	.80# 12.8 oz
250	2.5# 40.0#	.5# 8.0 oz	3.75# 60.0 oz	.75# 12.0 oz	5.0# 80.0 oz	1.0# 16.0 oz
300	3.0# 48.0 oz	.60# 9.6 oz	4.5# 72.0 oz	.90# 14.4 oz	6.0# 96 oz	1.2# 19.2 oz
350	3.5# 56.0 oz	.70# 11.2 oz	5.25# 84.0 oz	1.05# 16.8 oz	7.0# 112.0 oz	1.4# 22.4 oz
400	4.0# 64.0 oz	.80# 12.8 oz	6.0# 96.0 oz	1.2# 19.2 oz	8.0# 128.0 oz	1.6# 25.6 oz

Note: # = pound

Metric Table

Kgs of sand	1% Resin	Catalyst	1.5% Resin	Catalyst	2% Resin	Catalyst
23 kgs	.22 kgs	.04 kgs	.34 kgs	.06 kgs	.45 kgs	.09 kgs
45 kgs	.45 kgs	.09 kgs	.68 kgs	.13 kgs	.90 kgs	.18 kgs

Note: Most standard batch mullers do not efficiently mull less than 100 pounds of sand and do not mull more than 400–500 pounds of sand, though smaller mullers can be found. Clean the muller thoroughly with a wire brush following use. Some sculptors have found that a cement mixer with blades removed (and steel chunks and balls inserted) will work as a cheap muller.

Appendix D: Converting Cubic Inches to Sand Weight

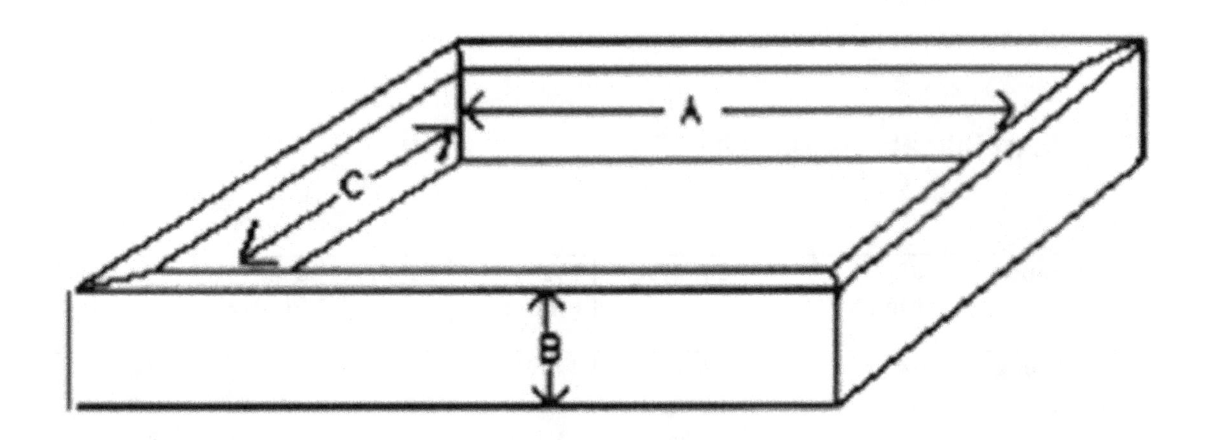

To find cubic inches or centimeters multiply inside dimensions of length (A) X height (B) X width (C)

432 cubic inches = 25# sand
864 cubic inches = 50# sand
1296 cubic inches = 75# sand
1728 cubic inches = 100# sand > **1 cubic foot**

2160 cubic inches = 125# sand
2592 cubic inches = 150# sand
3024 cubic inches = 175# sand
3456 cubic inches = 200# sand > **2 cubic feet**

3888 cubic inches = 225# sand
4320 cubic inches = 250# sand
4752 cubic inches = 275# sand
5184 cubic inches = 300# sand > **3 cubic feet**

5616 cubic inches = 325# sand
6048 cubic inches = 350# sand
6480 cubic inches = 375# sand
6912 cubic inches = 400# sand > **4 cubic feet**

Note: # = pound
Metric: a flask that is 30 cm. x 30 cm. x 30 cm. will weigh approximately 45 kgs

Appendix E: Sodium Silicate Binder Chart

Sodium Silicate Binder chart

# of Sand	**3% Binder**	**Catalyst**	**4% Binder**	**Catalyst**	**5% Binder**	**Catalyst**
100# or 45 kgs	3# or 48 Oz 1.36 kgs	.3# or 4.8 oz .14 kgs	4# or 64 oz 1.8 kgs	.4# or 6.4 oz .18 kgs	5# or 80 oz 2.26 kgs	.8# or 8.0 oz .36 kgs
200# or 90 kgs	6# or 96 oz 2.7 kgs	.6# or 9.6 oz .27 kgs	8# or 128 oz 3.6 kgs	.8# or 12.8 oz .36 kgs	10# or 160 oz 4.5 kgs	1# or 16 oz .45 kgs
300# or 136 kgs	9# or 144 oz 4.0 kgs	.9# or 14.4 oz .40 kgs	12# or 192 oz 5.4 kgs	1.2# or 19.2 oz .54 kgs	15# or 240 oz 6.8 kgs	1.5# or 21 oz .68 kgs

Note: #=pounds

APPENDIX F: Patina Formulas

Bronze Patinas

Basic Brown (good base coat patina)

ferric nitrate	one heaping teaspoon
hot water	one pint

Heat bronze to 200° F [93.3° C], apply solution with brush or spray; light to dark brown.

Basic Black (Liver of sulfer)

sulferated potash	one chunk the size of a little fingernail
hot water	one quart

Heat bronze to 200° F [93.3° C], apply solution with brush or spray; rinse frequently.

Basic Green

cupric nitrate	one heaping teaspoon
hot water	one pint

Heat bronze to 200° F [93.3° C], apply with brush or spray.

Antique Green

cupric sulphate	3 level teaspoons
ammonium chloride	1/2 teaspoon
water	one quart

Heat bronze to 200° F [93.3° C], apply with spray or brush.

Brown to Black

ammonium sulphide	1 teaspoon
water	one pint

Heat bronze to 200° F [93.3° C], apply with brush or spray. Repeated applications produce black; rinse frequently.

Yellow Green

ammonium chloride	2 teaspoons
cupric acetate	1 teaspoon
hot water 200° F [93.3° C]	1 pint

Brush or spray on clean surface.

Antique White

bismuth nitrate	2 teaspoons
hot water	1/2 pint

Heat bronze to 225° F [107° C];
spray or brush on clean surface. Solution should sizzle off bronze; keep solution suspended.

Verde

cupric nitrate	1 teaspoon
ammonium chloride	1 teaspoon
calcium chloride	1 teaspoon
water	1 quart

Heat bronze and apply with brush or spray; can also be used by immersion.

Japanese Foundation Brown

cupric sulphate	5 ounces
cupric acetate	5 ounces
water	1 gallon

Dip clean casting into solution at room temperature.

Japanese Foundation Green

cupric sulphate	5 ounces
cupric acetate	5 ounces
cupric carbonate	5 ounces
water	1 gallon

Dip clean casting into solution at room temperature.

Brown

ferric nitrate	2 ounces sodium
thiosulphate	2 ounces
water	1 pint

Brush or spray on clean surface.

Rust Red

ferric nitrate	1 teaspoon
ferric oxide	3 heaping teaspoons
hot water	1 quart

Mix, shake, allow to stand; shake prior to spray or brush application to hot metal.

White

zinc oxide	4 teaspoons
hot water	1 quart

Shake vigorously prior to application with spray or brush to bronze; bronze at 225° F [107° C].

Silver Gray

silver nitrate	1/2 teaspoon
water	1 pint

Spray or brush to bronze at 200° F [93.3° C].

Purple-Green

cupric nitrate	2 tablespoons
ammonia water	1 pint

Spray on hot bronze 225° F [107° C];
seal with wax immediately upon drying.

Iron Patinas

Black

caustic soda	3 teaspoons
potassium nitrate	1 teaspoon
boiling water	1 quart

Spray or brush to warm iron; **mix and use in well ventilated area only.**

Black

cupric sulphate	2.5 ounces
potassium chlorate	3 ounces
sodium chloride	3 ounces
boiling water	1 gallon

Fast acting; dip and move in solution until desired color is achieved.

Green

cupric chloride	3 ounces
ammonium chloride	1/2 ounce
boiling water	1 pint

Brush or spray on hot iron; seal with hot linseed oil after drying.

"Ohaguro" (traditional Japanese red brown)

0.15 liters	Water strained from boiling rice
1.50 liters	Sake
0.90 liters	Boiled water allowed to cool

Heat 4" piece of 1/2" iron rod till red hot. Drop in liquid mixture, cover tightly. Allow to age for 6 months. Apply to iron piece; applications of green tea will produce black.

Yellow-Orange to Rust Red

ferric nitrate	1/2 teaspoon
sodium chloride	1 teaspoon
water	1 quart

Spray or brush to clean surface; allow to dry, repeat until desired color is achieved.

Colors are more limited on iron (Red, Oranges, Yellows, Black, Blue, Purple White). Iron is an active oxygen-loving metal. Dry thoroughly and seal with wax. Some patinas may continue to "work" even after waxing. Monitor flaking and "rust." Some bronze patinas will work on iron as well, but often the result is not similar to what would happen on bronze. Greens are especially fugitive.

Aluminum Patinas

Black

cupric sulphate	8 ounces
zinc chloride	8 ounces
hydrochloric acid	1/2 ounce
warm water	

Dip in warm solution until desired color is achieved.

Blue

ferric chloride	3 teaspoons
ferric ferricyanide	3 teaspoons
water	1 quart

Heat to 150° F; apply with brush or spray.

Appendix G: Fluxing and Degassing Procedures

Aluminum fluxing and degassing

Find good commercial degassers and fluxes at www.foseco.com/. For the small studio, you will need to find a commercial foundry using these products and then negotiate the purchase of small amounts.

Coveral - An aluminum flux that minimizes oxidation, reduces gas absorbtion, and removes oxides from the melt. Produces a metal-free dross on the melts surfaces and facilitates skimming. For a 30 lb. melt, use 1/3 lb. of Coverall, added in two stages: 1/2 as soon as the initial charge begins to melt and then another 1/2 when the melt is complete. Measure each amount into a small piece of aluminum foil and seal. When adding the second amount, tum down the furnace to idle and then add the aluminum foil flux packet and wait till the flux has melted before resuming normal blast.

Copper alloy fluxing and degassing

Old time melters also used glass (beer bottle) or borax or sand on yellow brass or red brass. Melting with natural charcoal on the top of the melt will also control oxidation and gas absorption. Most silicon bronzes can be melted without cover fluxes. The following information is from the foseco.com website: http://www.foseco.com/en-gb/end-markets/foundry/products-services/non-ferrous-foundry/non-ferrous-foundry-detail/productsinfo/metal-treatment-2/treatment-products-for-copper-alloys/

Fluxes for cleaning, remelting, and element removal

CUPREX 1 tablets and RAFFINATOR 91 powder are cleaning agents with a vigorous oxidizing action for high conductivity or commercial copper, red brass and bronzes used in sand or die casting.

ELIMINALU 8 is a highly oxidizing agent for cleaning and removing aluminium contamination from copper alloys. It is not suitable for aluminium bronzes and only to a certain extent for manganese or silicium bronzes.

Covering fluxes

CUPREX Standard (powder), CUPREX 100 (powder), and CUPREX 14 BL (block) are covering agents for copper alloys that provide the melt with oxidizing conditions. The products can be used for copper alloys containing tin and nickel such as red brass, tin bronze, lead bronze (with less than 10% lead), phosphor bronze and copper-nickel alloys such as Monel. Use 1 lb. for a 90 lb. melt.

CUPRIT 8 and CUPRIT 49 are covering agents in powder form for copper and copper alloys melted under neutral conditions. They are suitable for high-conductivity copper (CUPRIT 8), brasses, special brasses and soldering alloys (CUPRIT 49). Use 1 lb. for a 90 lb. melt

CHROMBRAL 4 is a covering agent used in the melting of chrome bronzes and other chrome-containing copper alloys. It is also suitable for chrome-alloyed, highly conductive copper.

ALBRAL are reducing cleaning and covering fluxes for aluminum bronzes and special brasses containing alloying elements that are prone to oxidation such as aluminum, manganese, and silicon. ALBRAL 2 produces a liquid dross on the metal surface while ALBRAL 3 produces a dry dross. For a 90 lb. melt, 1/4 lb. is added with the metal and an additional 1/4 lb. is added 15 minutes before pouring.

Purging/degassing and homogenizing copper melts

LOGAS 50 briquettes are used for plunging into copper and copper alloys to remove dissolved hydrogen and float out oxides. For a 90 lb. melt, plunge 2 oz. into the melt and hold till reaction subsides.

DEOX TUBES are individual copper tubes containing a variety of reagents and are specially designed to remove dissolved oxygen formed during the melting process.

RECUPEX 120 is a neutral cleaning powder for brass used in sand casting and is particularly suitable for use with scrap which has been contaminated with oil. ELIMINALU 8 is a highly oxidizing agent for cleaning and removing aluminium contamination from copper alloys. It is not suitable for aluminium bronzes and only to a certain extent for manganese or silicium bronzes.

Covering fluxes

CUPREX Standard (powder), CUPREX 100 (powder), and CUPREX 14 BL (block) are covering agents for copper alloys that provide the melt with oxidizing conditions. The products can be used for copper alloys containing tin and nickel such as red brass, tin bronze, lead bronze (with less than 10% lead), phosphor bronze and copper-nickel alloys such as Monel. Use 1 lb.for a 90 lb. melt.

CUPRIT 8 and CUPRIT 49 are covering agents in powder form for copper and copper alloys melted under neutral conditions. They are suitable for high-conductivity copper (CUPRIT 8), brasses, special brasses and soldering alloys (CUPRIT 49). Use 1 lb. for a 90 lb. melt

CHROMBRAL 4 is a covering agent used in the melting of chrome bronzes and other chrome-containing copper alloys. It is also suitable for chrome-alloyed, highly conductive copper.

ALBRAL are reducing cleaning and covering fluxes for aluminum bronzes and special brasses containing alloying elements that are prone to oxidation such as aluminum, manganese, and silicon. ALBRAL 2 produces a liquid dross on the metal surface while ALBRAL 3 produces a dry dross. For a 90 lb. melt, 1/4 lb. is added with the metal and an additional 1/4 lb. is added 15 minutes before pouring.

Purging/degassing and homogenizing copper melts

LOGAS 50 briquettes are used for plunging into copper and copper alloys to remove dissolved hydrogen and float out oxides. For a 90 lb. melt, plunge 2 oz. into the melt and hold till reaction subsides.

DEOX TUBES are individual copper tubes containing a variety of reagents and are specially designed to remove dissolved oxygen formed during the melting process. Tubes are available for all types of copper-based alloys; thermal and electrical conductivity of the alloys are not affected.

SLAX 20 is a coagulant used for positive slag control on all copper alloys.

Iron fluxes and slag coagulants

Limestone is the traditional flux for iron in a cupola or cupolette. Perlite (sometimes called "popcorn") or other slag coagulants (such as Slax or Slag Grip) are used on the top of the ladle when the metal is near pouring temperature. They form a thick chunk of slag. Removal is easier when all the slag holds together.

Appendix H: Cupola

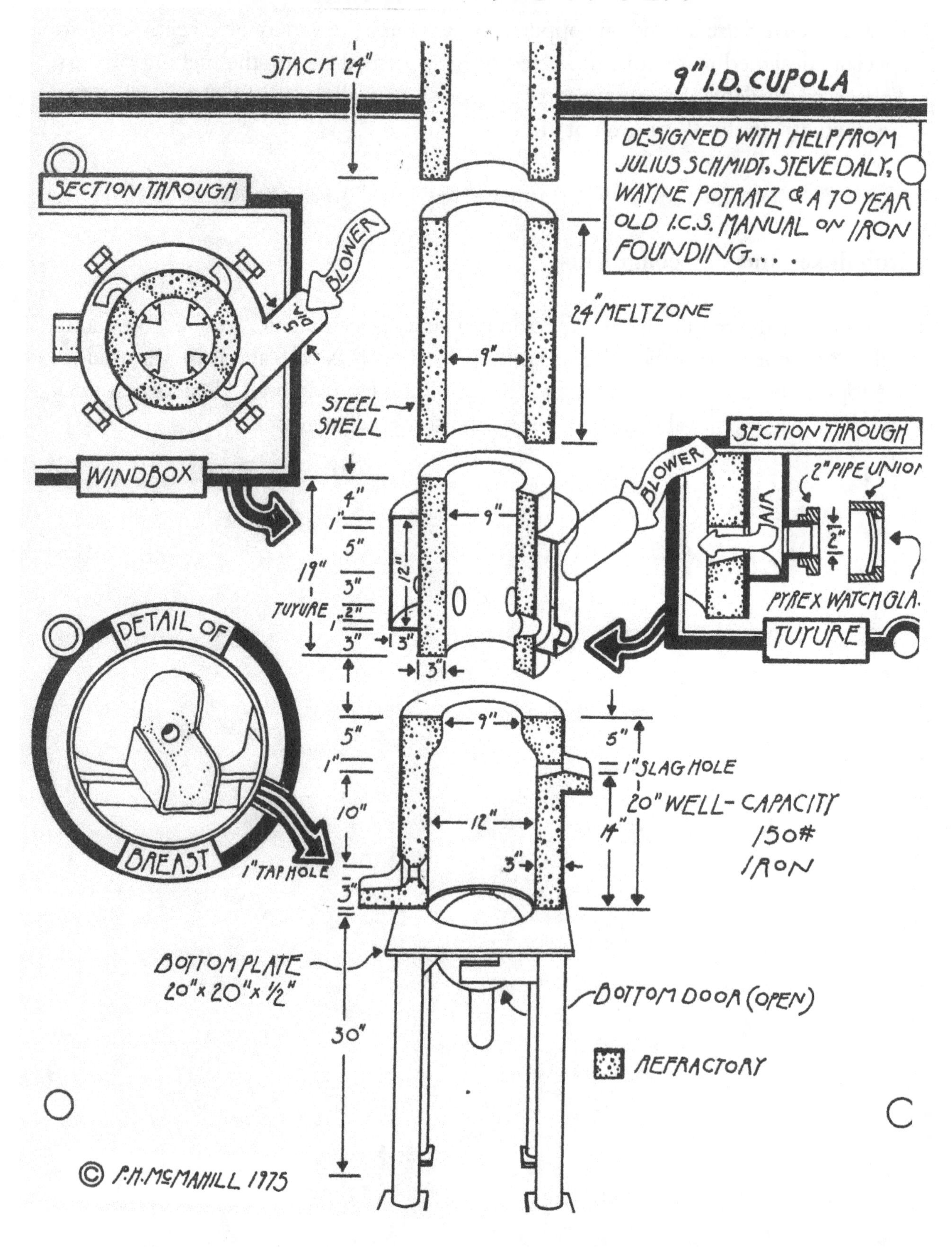

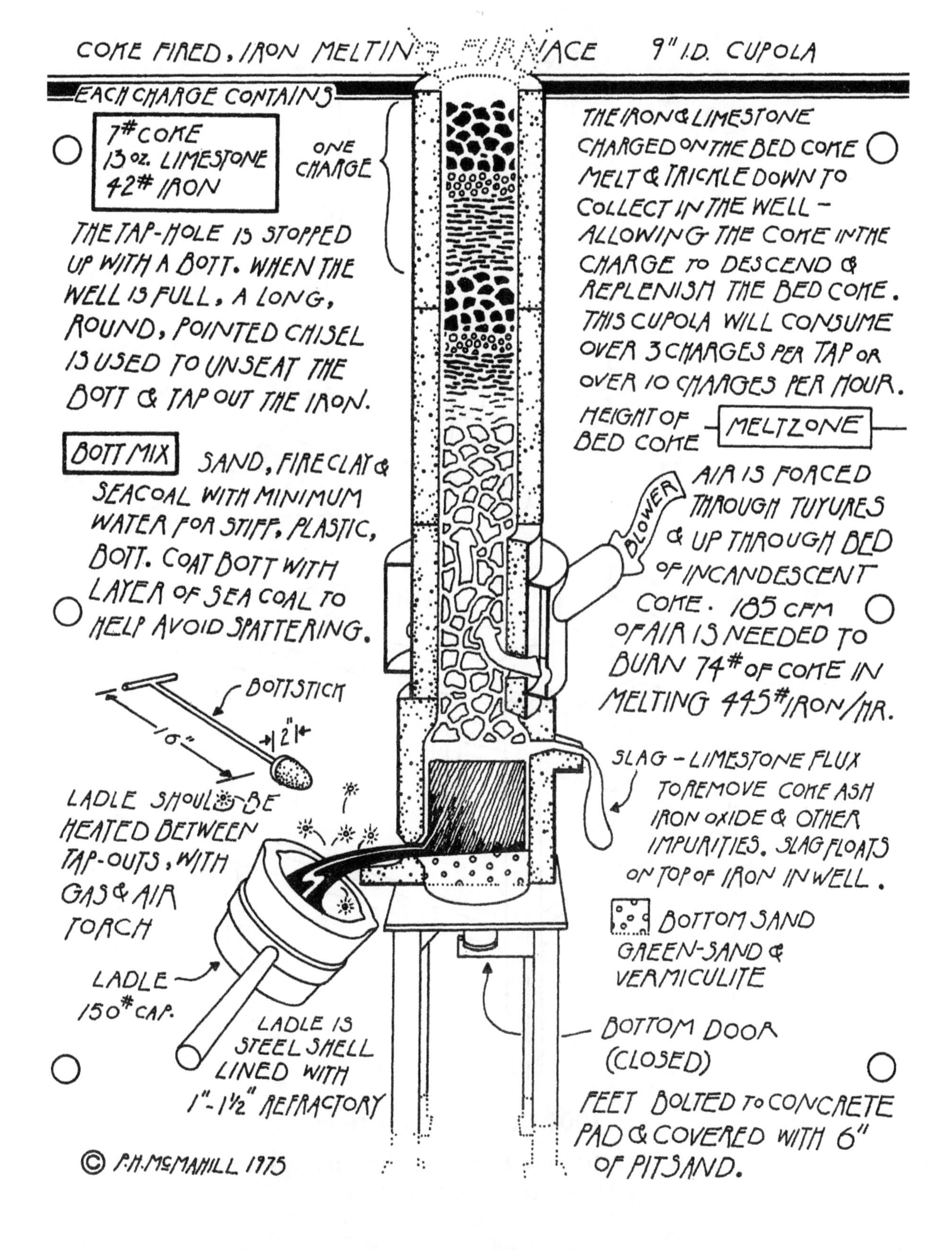
COKE FIRED, IRON MELTING FURNACE 9" I.D. CUPOLA
EACH CHARGE CONTAINS
7# COKE
13 oz. LIMESTONE
42# IRON
ONE CHARGE
THE TAP-HOLE IS STOPPED UP WITH A BOTT. WHEN THE WELL IS FULL, A LONG, ROUND, POINTED CHISEL IS USED TO UNSEAT THE BOTT & TAP OUT THE IRON.
BOTT MIX
SAND, FIRECLAY & SEACOAL WITH MINIMUM WATER FOR STIFF, PLASTIC, BOTT. COAT BOTT WITH LAYER OF SEA COAL TO HELP AVOID SPATTERING.
BOTTSTICK
10"
2"
LADLE SHOULD BE HEATED BETWEEN TAP-OUTS, WITH GAS & AIR TORCH
LADLE 150# CAP.
LADLE IS STEEL SHELL LINED WITH 1"-1½" REFRACTORY
© P.H. McMAHILL 1975
THE IRON & LIMESTONE CHARGED ON THE BED COKE MELT & TRICKLE DOWN TO COLLECT IN THE WELL - ALLOWING THE COKE IN THE CHARGE TO DESCEND & REPLENISH THE BED COKE. THIS CUPOLA WILL CONSUME OVER 3 CHARGES PER TAP OR OVER 10 CHARGES PER HOUR.
HEIGHT OF BED COKE
MELTZONE
BLOWER
AIR IS FORCED THROUGH TUYURES & UP THROUGH BED OF INCANDESCENT COKE. 185 CFM OF AIR IS NEEDED TO BURN 74# OF COKE IN MELTING 445# IRON/HR.
SLAG - LIMESTONE FLUX TO REMOVE COKE ASH IRON OXIDE & OTHER IMPURITIES. SLAG FLOATS ON TOP OF IRON IN WELL.
BOTTOM SAND GREEN-SAND & VERMICULITE
BOTTOM DOOR (CLOSED)
FEET BOLTED TO CONCRETE PAD & COVERED WITH 6" OF PITSAND.

Appendix I: Cupolette

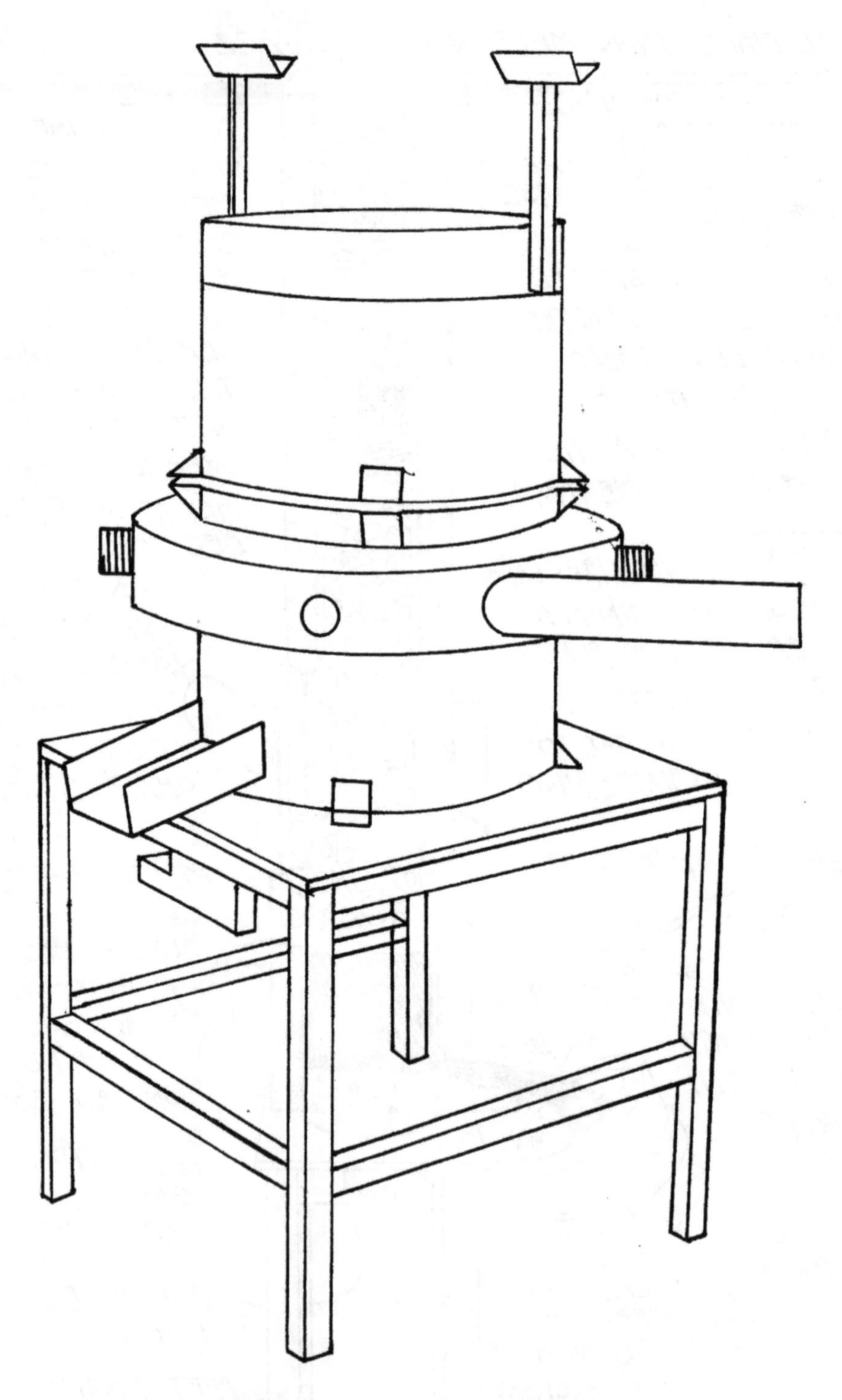

100# Cupolette Furnace

Source: Used with permission from Professor Kurt Dyrhaug, Lamar University

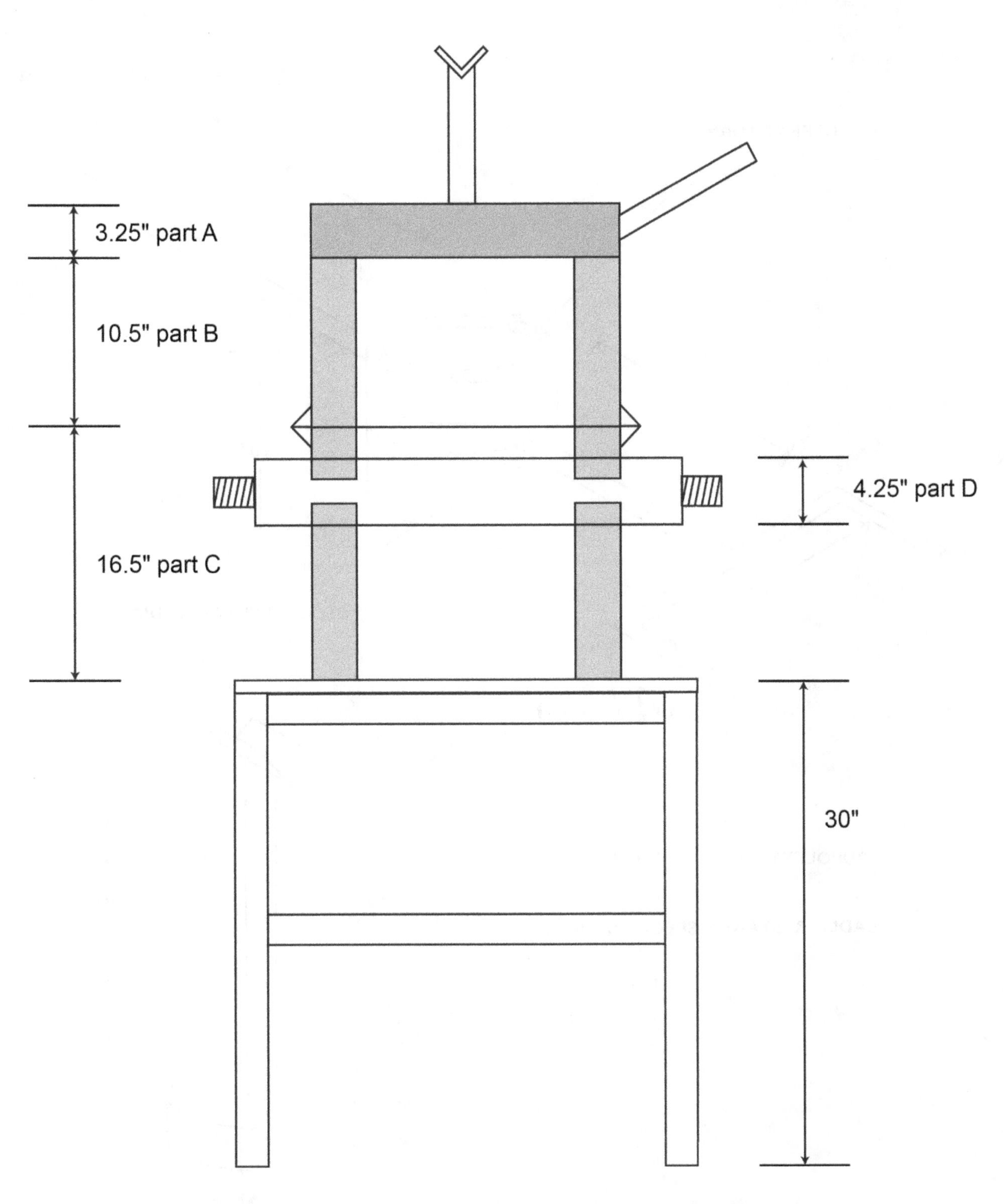

100# Cupolette Furnace Specifications

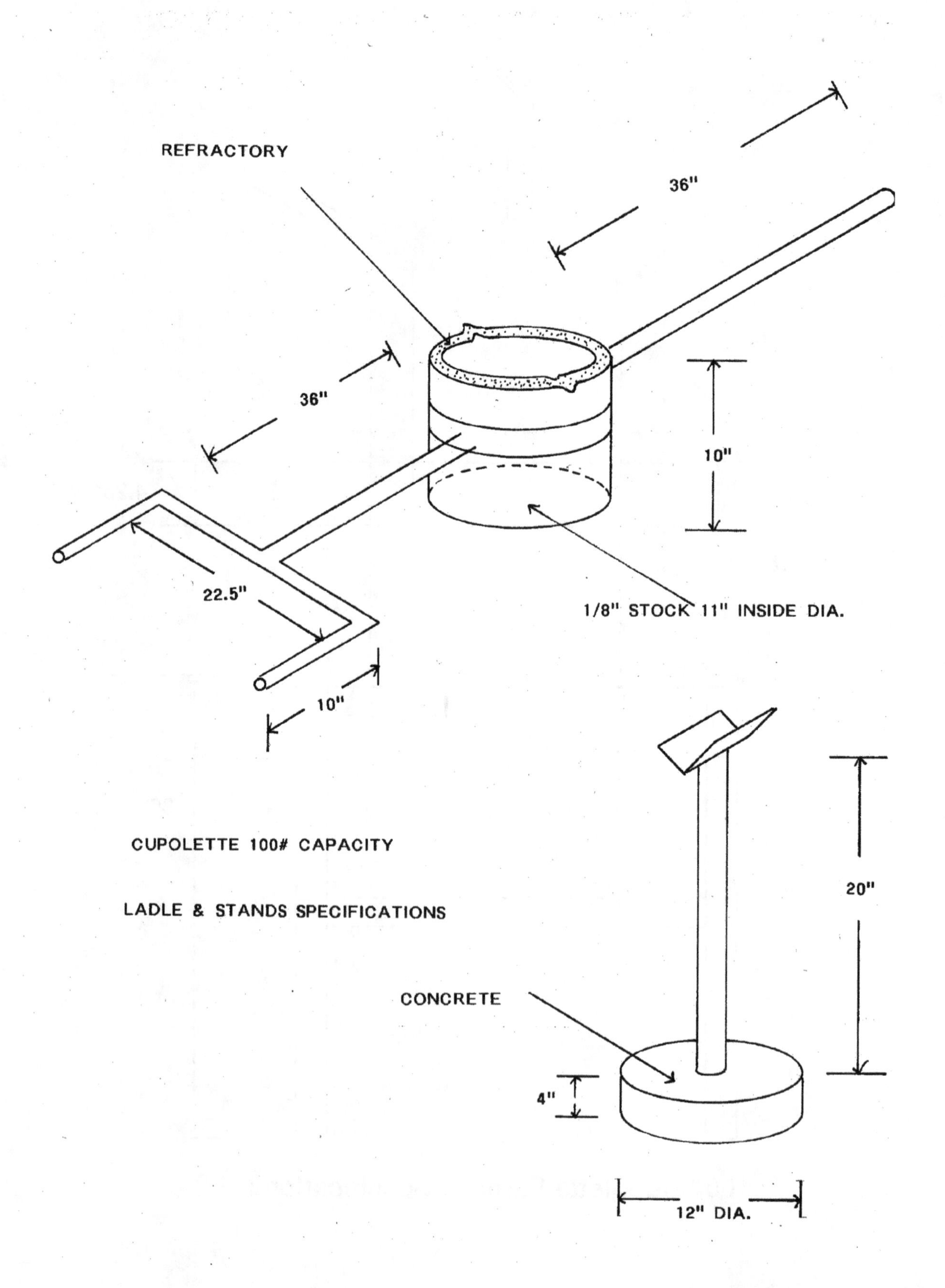
REFRACTORY
36"
36"
10"
22.5"
1/8" STOCK 11" INSIDE DIA.
10"
CUPOLETTE 100# CAPACITY
LADLE & STANDS SPECIFICATIONS
20"
CONCRETE
4"
12" DIA.

Iron Furnace Dimensions
(Stock: 1/8" mild steel)

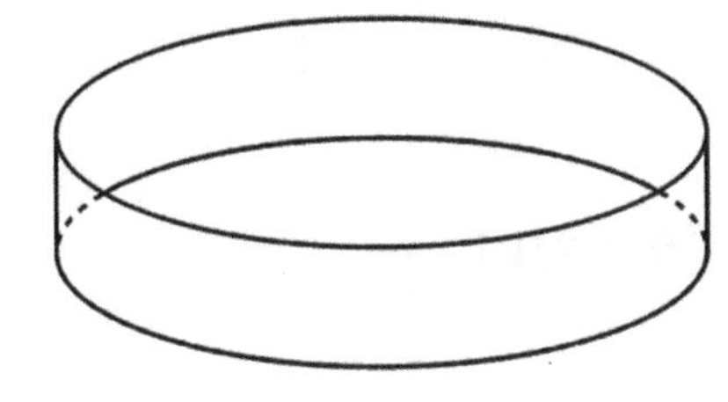

Part A
Quantity: 1
3.5" (h) x 20" (O.D.)

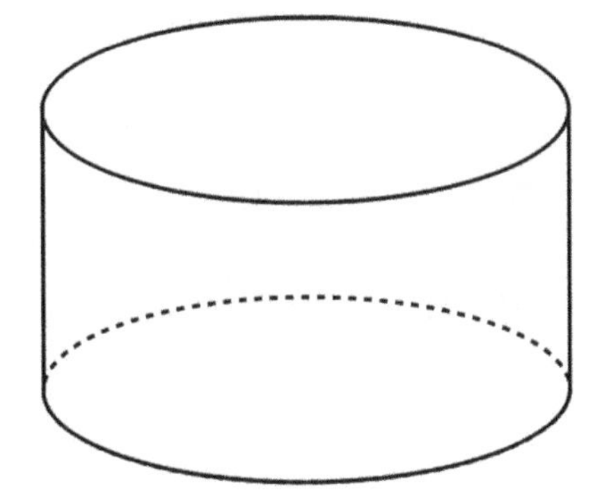

Part B
Quantity: 1
10.5" (h) x 20" (O.D.)

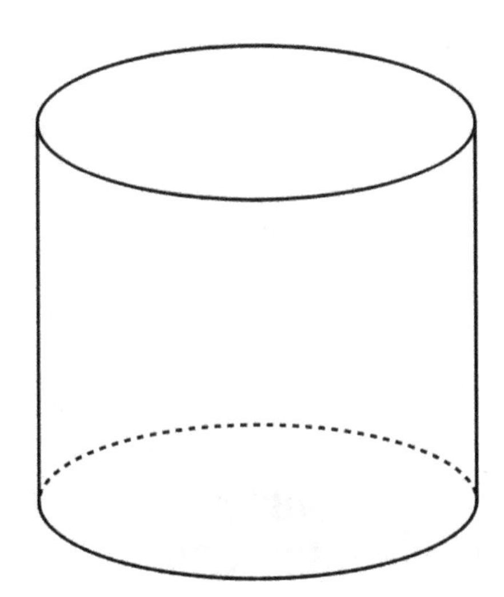

Part C
Quantity: 1
16.5" (h) x 20" (O.D.)

Source: Used with permission from Professor Kurt Dyrhaug, Lamar University

Iron Furnace Dimensions
(Stock: 1/8" mild steel)

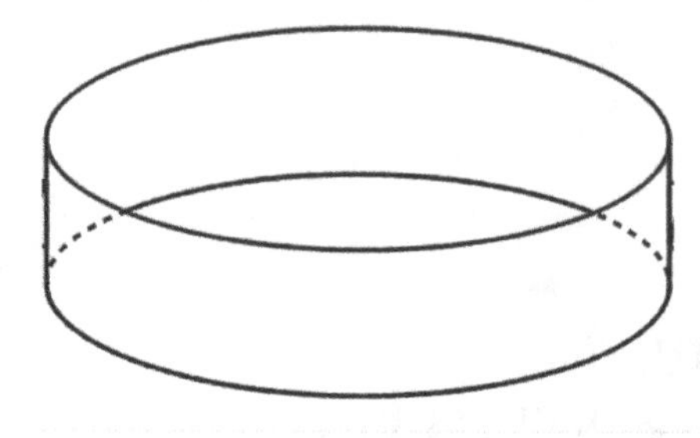

Part D
Quantity: 1
4.25" (h) x 28" (I.D.)

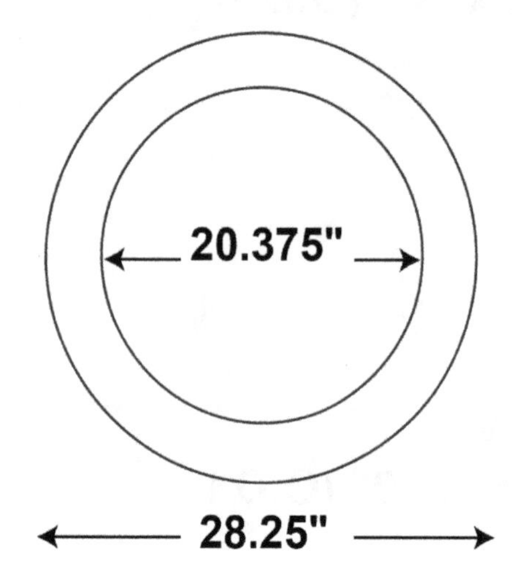

Part E
***Quantity: 2**
10.5" (h) x 20" (O.D.)

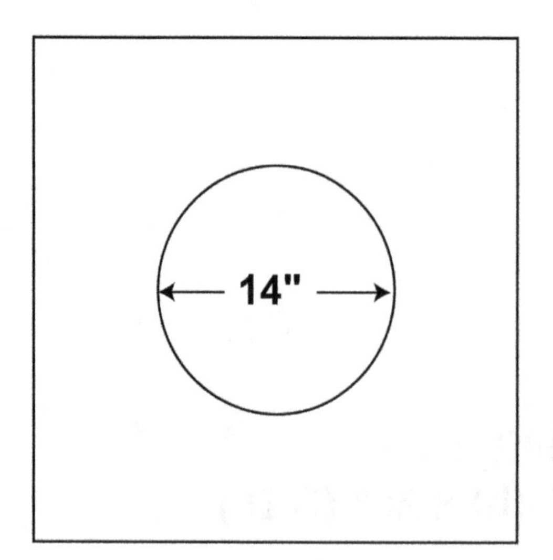

Need 14" cut out

Part F
Quantity: 1
***Stock: 3/8" mild steel**
Dimensions: 30" x 30"

Source: Used with permission from Professor Kurt Dyrhaug, Lamar University

Appendix J: Cupola Design Information

Cupola Design Information

D= Inside diameter of melt-zone
Approximate capacity of cupola lbs./ hr.)= .7854 D2 x 7
Total tuyere area at refractory shell= .7854 D2/4

Air required through tuyeres= 150 Cu. Ft. per lb. of Coke/ Hr.
CF/M Air through tuyeres= (.7854 02 x 7) x 150/60
Maximum blast pressure= 4 oz.
Suggested blower CF/M= .7854 D2 x 2.5

Bed height above tuyeres (in inches)= 10.5 $\sqrt{\text{Blast Pressure} + 6}$
Well capacity (lb. of iron)= .102 D2 x C (C= distance from tap hole to slag hole)

Metal charge (lb. of iron)= .78854 D2/1.5
Coke size= 1/10 to 1/12 D

Coke charge = Metal/ Coke = 6/1

Limestone= 0.175 to 0.225 lb. of limestone/ 1 lb. of metal charge or 1.75 lb. [80 kilos] 100 lb. [kilos] of metal charge

Blower Specs for 100# Cupolette

Blower, Motor Power Rating 1 HP, Phase Single, Wheel Diameter 10 9/16 Inches, Radial Blade, With Motor

Grainger Item #	7C447
Price (ea.)	**$524.00**
Brand	DAYTON
Mfr. Model #	7C447
Ship Qty.	1
Sell Qty. (Will-Call)	1
Ship Weight (lbs.)	53.0
Availability	**Ready to Ship**
Catalog Page No.	4332
Country of Origin **(Country of Origin is subject to change.**	USA

Appendix K: Iron Pour Record Sheet

45th Annual Minnesota Iron Pour, Friday April 25, 2014

Fire Lit: 8:00 am
Burn @Slagger: 8:45 am
Burn @ Tuyeres: 9:00 am
Burned in: 9:50 am
Air on ¼: 9:50 am
Torch on Big Ladle: 9:50 am
Total weight poured: 1650#

Air on Full: 11:00 am
First Charge: 11:09 am
First Drops: 11:10 am
Fast Drops/ Botted up: 11:11 am
Total # of molds: 63
Bottom drop/ Bagpipe Solo: 2:00 pm
Salute to the Pour: 2:15 pm

Charge	Wt.	Charge Time	Tap Time	Time Twixt	Temp	Notes
1	250#	11:09	11:27	18 M	2343F	Tapper: Tobias Flores 1% O2
2	250#	11:30	11:50	20	2359	Tapper: Thomas Gipe 1% O2
3	250#	11:53	12:16	23	2391	Tapper: Kurt Dyrhaug 1% O2
4	250#	12:20	12:40	20	2383	Tapper Cliff Prokop 1% O2
5	250#	12:44	1:04	18	2424	Tapper: Jenny Nellis; shorter tap time, hotter iron. Suggest shorter tap time when doing 250#
6	250#	1:08	1:28	20	2398	Tapper: Bill Malo
7	150#	1:30	1:45	25	2397	Tapper: Chris Larson

Note: # = pounds

Guest Artists:
Professor Kurt Dyrhaug, Lamar University, Beaumont TX
Professor Tobias Flores, Fort Hays State University, Hays KS
Professor Emeritus Thomas Gipe, SIUE, Southern Illinois U, Edwardsville
Professor Emeritus Jenny Nellis, U of MN Morris
Professor Gary Wahl, U of MN, Morris, MN
Professor Cliff Prokop, Keystone College, La Plume, PA
Professor Emeritus Jim Swartz, S.W. State University, Marshall, MN
Professor Tamsie Ringler, St. Catherine's University, MN
Professor Irve Dell and students from St. Olaf College, Northfield, MN
Professor David Lobdell, New Mexico Highlands University, Las Vegas, NM
Do Kyun Lee, Gwanju, South Korea
Bill Malo, Martinez, CA
Kelly Ludeking, Decorah, Iowa
Tom Christiansen, Lutsen, MN
Melanie Van Houten, Josephine Sculpture Park, Frankfort, KY
Wayne E. Potratz, Minneapolis, MN

INDEX